The New Cosmos
Answering Astronomy's Big Questions

Over the past decade, astronomers, planetary scientists, and cosmologists have answered – or are closing in on the answers to – some of the biggest questions about the universe. David J. Eicher presents a spectacular exploration of the cosmos that provides you with a balanced and precise view of the latest discoveries. Detailed and entertaining narratives on compelling topics such as how the Sun will die, the end of life on Earth, why Venus turned itself inside-out, the Big Bang Theory, the mysteries of dark matter and dark energy, and the meaning of life in the universe are supported by numerous color illustrations, including photos, maps, and explanatory diagrams. In each chapter, the author sets out the scientific history of a specific question or problem, before tracing the modern observations and evidence in order to solve it. Join David J. Eicher on this fascinating journey through the cosmos!

DAVID J. EICHER is Editor-in-Chief of *Astronomy* magazine, and one of the most recognized astronomy enthusiasts in the world. He has spoken widely to amateur astronomy groups and written eight books on astronomy, including *Comets: Visitors from Deep Space* and *The Universe from Your Backyard*. He is president of the Astronomy Foundation and a member of the Board of Directors for Starmus Festival. An avid observer of astronomical objects for more than 35 years, he was honored by the International Astronomical Union with the naming of a minor planet, 3617 Eicher.

The New Cosmos

Answering Astronomy's Big Questions

David J. Eicher

Editor-in-Chief, *Astronomy* magazine

Foreword by Alex Filippenko

University of California, Berkeley

CAMBRIDGE
UNIVERSITY PRESS

University Printing House, Cambridge CB2 8BS, United Kingdom

Cambridge University Press is part of the University of Cambridge.

It furthers the University's mission by disseminating knowledge in the pursuit of education, learning, and research at the highest international levels of excellence.

www.cambridge.org
Information on this title: www.cambridge.org/9781107068858

First published 2015

Printed in the United Kingdom by Bell and Bain Ltd

A catalog record for this publication is available from the British Library

Library of Congress Cataloging in Publication Data
Eicher, David J., 1961-
The new cosmos : answering astronomy's big questions / David J. Eicher, editor-in-chief, *Astronomy* magazine ; foreword by Alex Filippenko, University of California, Berkeley.
     pages   cm
Includes bibliographical references and index.
ISBN 978-1-107-06885-8 (Hardback)
1. Astronomy.   2. Cosmology.   I. Title.
QB43.3.E33 2015
520–dc23   2015023761

ISBN 978-1-107-06885-8 Hardback

Now that anyone is free to print whatever they wish, they often disregard that which is best and instead write, merely for the sake of entertainment, what would be best forgotten, or, better still be erased from all books. And even when they write something worthwhile they twist and corrupt it to the point where it would be much better to do without such books, rather than having a thousand copies spreading falsehoods over the whole world.

– Niccolò Perotti (1429/30–1480)

If the Lord Almighty had consulted me before embarking on creation thus, I should have recommended something simpler.

– Alfonso X of Castile (1221–1284)

Contents

Foreword

It's often said that we now live in a "golden age" of astronomy – and indeed, it's true.

Hardly a week goes by without a press release informing the public of a new cosmic discovery made with one or more instruments from the existing arsenal of ground-based and space-based telescopes, or through theoretical and computational studies. Sometimes, especially during national and international meetings such as those of the American Astronomical Society, several new celestial findings are announced each day.

And what amazing developments we have witnessed! Just 2 decades ago, in the mid-1990s, the evidence for "dark matter" was tantalizing but not compelling, contrary to the present situation. In a still greater turn of events, hardly anyone considered the possibility that even more mysterious "dark energy" dominates the mass-energy content of the universe and accelerates its expansion, probably dooming us to end in the Big Chill. Only the first few exoplanets had been discovered, foreshadowing the thousands that are now known or strongly suspected, and the detection of Earth-like exoplanets in the near future was considered unlikely. Given the few observed objects in the Kuiper Belt and their relatively small sizes, for another decade most astronomers would still consider Pluto to be a genuine planet. Black holes, previously popular mainly among the fertile minds of theoretical physicists and science-fiction writers, were suspected to exist but not yet detected beyond reasonable doubt.

As a long-time editor of the well-known *Astronomy* magazine, author David Eicher gained much experience in bringing celestial highlights to amateur astronomers and astronomy enthusiasts. One could learn the basics of astronomy and physics in standard textbooks, such as my own *The Cosmos: Astronomy in the New Millennium* (coauthored with Jay M. Pasachoff), but such books are not able to describe new discoveries on short timescales, and much of their volume must remain devoted to the fundamentals. Conversely, amateur astronomy periodicals generally do not provide very detailed analyses of the state of various astronomical subfields, concentrating instead on relatively brief summaries of recent developments.

In this book, on the other hand, Eicher synthesizes a wealth of relatively new information into an interesting, coherent, up-to-date overview of many of the most important and exciting areas of astronomy. Inspired as a youngster by the legendary Carl Sagan, and having a deep love of bringing the cosmos to the general public, he passionately shares his broad knowledge and presents the state of the art on topics ranging from our own solar system and the Milky Way Galaxy, to other planetary systems and galaxies, to the lives and deaths of stars, and to the overall composition and fate of the cosmos. Along the way, he considers the future of life on Earth, the possibility of life elsewhere, and even the meaning of life.

Two decades from now, it will be interesting to see how much our view of the universe will differ from what is presented here. Will we better understand the origin of the Moon, the absence of water on Mars, and the geologic history of Venus? Will we have mapped oceans and continents on exoplanets in the Goldilocks zone? Will we have directly detected exotic particles thought to be responsible for most of the dark matter in the cosmos? Will we know the true nature of dark energy and the ultimate fate of the universe?

Immersing yourself in the chapters that follow will give you a new sense of wonder at the amazing universe that we inhabit, yet simultaneously awakening both your humility as a cosmic speck of dust and your awe at our present understanding of the cosmos. And there is little doubt that this book will help inspire inquiring minds to further investigate the universe and its contents, as the human brain is restless and there is still far more to be learned.

Alex Filippenko
University of California, Berkeley

Preface

I was a child of *Cosmos*.

My youth seemed connected to Carl Sagan. When I was 14, I attended my first "star party" by accident, catching a glimpse of Saturn and other attractions in a small reflecting telescope, and that moment changed the world for me. I became active in the local astronomy club in Oxford, Ohio, a small university town where my father was a professor of organic chemistry at Miami University. The local club needed a writer on deep-sky objects – star clusters, nebulae, and galaxies – and recruited me. Soon I was so entranced with writing about these mysterious creatures of the universe beyond our solar system that I started an amateur publication, *Deep Sky Monthly*, that had its genesis on the mimeograph machine in my father's chemistry office. It was the summer of 1977, and I was 2 months shy of 16.

During the first months of producing a publication for astronomy enthusiasts, while in high school, I wrote Professor Carl Sagan at Cornell University, letting him know about the publication and seeking career advice. He very graciously replied with the first of a number of letters. This was during his time as a celebrated astronomy figure – he periodically shared enthusiasm with Johnny Carson on *The Tonight Show* – but before his production of the legendary *Cosmos* TV program.

On June 6, 1977, Carl wrote me his first letter. His wisdom, encouragement, generosity, and positive spirit during every encounter we had from that moment on were a major factor in my pursuit of astronomy. "I am delighted to hear from a 15-year-old who is already so active in astronomy," he wrote, and after paragraphs of advice, he closed with "With all good wishes on your career."

My admiration for Carl Sagan grew throughout our correspondence and I beamed with pride in knowing Carl during the airing of his *Cosmos* series on PBS TV in 1980. The show premiered on Sunday, September 28, 1980, and I rushed inside after a busy day, a pleasant 72 °F in Oxford, to turn on the TV just in time for that haunting theme music by Vangelis.

Figure 0.1 David Eicher in June 1982, 5 years after commencing publication of his amateur-produced *Deep Sky Monthly* magazine, and several years after coming under the influence of Carl Sagan, at the Eicher House in Oxford, Ohio.
David J. Eicher

"The cosmos is all there is, or ever was, or ever will be," said Carl in his opening sequence. "Our contemplations of the cosmos stir us. There is a tingling in the spine, a catch in the voice, a faint sensation as if a distant memory of falling from a great height. We know we are approaching the grandest of mysteries."

I was immediately entranced by the series, as were millions. It became one of the great success stories for PBS in an era when relatively few TV channels existed – it propelled an entire generation to discover astronomy, the stars, and the universe around us. I was particularly proud when I received a copy of the Random House *Cosmos* book in the mail, inscribed to me by Carl – "For Dave Eicher / friend of the Cosmos."

The sensational momentum of enthusiasm in my life stoked by Carl Sagan never left me. After Miami University, at age 21, I came to Milwaukee to join the staff of *Astronomy* magazine, the world's largest publication on the subject, as its most junior assistant editor, and I brought my little magazine with me, now titled *Deep Sky* and published as a quarterly. I've been at *Astronomy* for more than 32 years, and have held every editorial job, becoming the magazine's editor-in-chief in 2002. We published *Deep Sky* for 10 more years, ending it in 1992 when it became clear that if I were to progress further with *Astronomy*, I couldn't also do the smaller magazine a day or so every week.

Figure 0.2 Carl Sagan inspired a generation with books and television programs that made science accessible through his brimming enthusiasm. The astronomer's best-known work, *Cosmos: A Personal Voyage,* ultimately reached half a billion people across the world, and won an Emmy Award.
NASA/JPL-Caltech

My life at *Astronomy* magazine has been great fun. But I've never forgotten the principles laid down in those letters and talks with Carl Sagan. A little more reminiscence about him, from a short piece I wrote for *Astronomy:*

My first encounter with Carl Sagan came as a teenager when I readied for a late night of observing with my telescope. I hovered over The Tonight Show *before heading outside and sat, mesmerized, as this Cornell University professor made astronomy relevant, exciting, and meaningful to ordinary Americans. Sagan didn't utter "billions and billions" on that summer night; that phrase, gently lampooning his friend Sagan, arrived later from amateur astronomer Johnny Carson.*

As I looked toward a career in astronomy and started publishing Deep Sky Monthly, *I sent a letter to Sagan and included copies of my little publication on observing galaxies, clusters, and nebulae. It was mid-1977. I was 15 and he was 42. Busy as he was (although this was pre-Cosmos), Sagan wrote me a long letter that I received a week or two later. It absolutely made my year and inspired me to redouble my efforts in astronomy. Filled with career advice, the letter served as an inspiration for years to come.*

Through a correspondence, I got to know Carl Sagan. What struck me most was how generous he was with his time and with his wisdom. He was truly a humanist who cared about people and was impressed with helping them in any way he could. When Cosmos ignited interest in astronomy through the airwaves of PBS, he sent me an inscribed copy of the accompanying Random House book, signed "For Dave Eicher/friend of the Cosmos." I was awestruck.

In those days, Carl Sagan took heavy criticism from conservative or even jealous professional colleagues over his popularization of astronomy and science. To some, it was "selling out to the masses." In the world that now exists, where too few people value science and know about its details, how those elitists would sing a different tune! Sagan realized the value of the Average Joe understanding and supporting science, and did more than just about anyone else to deepen that connection between science and the public.

When I finally met Carl Sagan and spent time with him at meetings, now as an editor at Astronomy, I was even more impressed. I was struck by his height (he was 5'11" but for whatever reason seemed taller than that in person and shorter on TV), his soft-spoken, fatherly manner, his fine manner of dress, and his patient, caring gaze.

I recall how impressive it was to hear his views, in person, on a wider variety of subjects. This was no shallow TV presenter. At the "Comets and the Origin and Evolution of Life" Meeting in Eau Claire, Wisconsin, in the fall of 1991, I sat with him in the back of a meeting room. He rattled off numerous interesting questions following many of the lectures, as well versed in a whole maze of specialized research as the speakers were.

Carl Sagan wrote for Astronomy magazine many times, from the earliest issues onward. He sent me his last contribution to the magazine in 1993, and it appeared in the 20th anniversary issue of the magazine, in a story about what was coming in astronomy's future.

Three years later, we all learned he was being treated for an illness, and I called him in late 1996 when he was in Seattle at the Hutchinson Cancer Research Center. He spoke, as always, with great caring in his voice. I asked him if he would be a member of the magazine's newly forming Editorial Advisory Board, and talked about all the exciting things to come, unaware of the grave seriousness of his illness. In the most generous, compassionate way you could imagine, he simply said, "Dave, nothing would make me happier."

Two weeks later, he was gone, and we were all left stunned. Astronomy had lost its best friend and would never be quite the same again.

One of Carl's best lines from *Cosmos* was: "The cosmos is full beyond measure of elegant truths, of exquisite interrelationships, of the awesome machinery of nature."

Cosmos was produced some 35 years ago, a very short span in the history of the universe, and yet human understanding of the cosmos has changed almost unbelievably since then.

In 1980, we knew nothing of dark energy, and far less than we do now about dark matter. The nature of black holes was still mostly conjecture. We did not know the intricate details of how the Sun will die, or what will happen when life comes to an end on Earth. We did not yet have a good idea of how the Moon formed. We utterly lacked an understanding of the nature of martian or venusian geological history. Pluto was still safely considered a planet. We had no evidence for numerous planets orbiting stars near us in the galaxy.

We erroneously believed the Milky Way was a normal spiral galaxy. We did not know about the future collision of the Milky Way and Andromeda galaxies.

We lacked detailed evidence for the exact nature of the Big Bang and the cosmic microwave background radiation (beyond its mere detection). We had only hazy notions of the universe's size, age, and fate. And we did not know enough about the origin of life on Earth or the number of stars in the universe to speculate meaningfully about the potential existence of life elsewhere in the cosmos.

All of these major areas were in a very primitive state just 35 years ago, when *Cosmos* first aired, and thus the reason for putting all this stuff together in a big book like *The New Cosmos*.

From the earliest days, as humans looked skyward, they no doubt wondered about the twinkling lights spread across the sky. For thousands of years, myths and imaginations carried the day as culture slowly leaned forward toward rational observation and empirical science. Astrologers dreamed up ideas; philosophers imagined answers. And then, in the autumn of 1609, Galileo climbed to the roof of his house in Padua, Italy, and – after looking at the steeple of the nearby church with his new telescope – slid the field of view over to the Moon. He became the first to see and share with the world a magnified view of the lunar surface, with craters, dark "seas," and mountains. In making this first shared telescopic observation, Galileo ignited a new era of astronomy.

For centuries thereafter, astronomy was, like the other sciences, mainly an exercise in classification. Whether they contemplated stars, minerals, butterflies, plants, or fish, scientists categorized what they observed in nature as a first gross step in understanding it. Then, in nineteenth-century astronomy, a revolution in astrophysics arrived. Photography, spectroscopy, and other tools led to a new wave of understanding the cosmos. Suddenly, astronomers began to understand the physical nature of the universe in a more enlightened way.

But another revolution has taken place, one that is far more fast-paced and remarkable. Over just the past decade, astronomers, planetary scientists, and cosmologists have answered, or are closing in on the answers to, some of the biggest questions about the universe.

This revolution is rapidly recasting what we know about the cosmos around us. Given this flood of new findings and new ways to understand nature, we can now say that we live in a place redefined by our fresh knowledge – a new cosmos.

These include questions about the universe's origin, its fate, its size, its shape, and its age. We now know about the relative numbers of planets around us in nearby parts of the Milky Way Galaxy. We know about the barred spiral structure of the Milky Way and about its future collision and merger with our neighbor in Andromeda. We know about the ubiquitous nature of black holes in galaxies. We know about the cosmic distance scale of the galaxy, about the vast stretches of space that surround us inside the Milky Way and beyond it.

Planetary scientists also pursue answers to big questions that are hotly evolving. These include such key questions as the formation of the Moon, the transformation of Mars' climate, the strange case of Venus turning itself inside out, arguments over what makes a planet and what is merely a dwarf planet. We also know what will happen to the Sun, how our solar system will look 5 billion years from now, and how long life will be able to exist on Earth before the oceans will boil away, ending our existence.

Answers to questions of cosmology lag behind a bit, given their enormous complexity, but astronomers have made huge recent strides that hint toward resolution. Studies are narrowing the gap on questions about the nature of dark matter, dark energy, the fate of the universe, the abundance of life in the universe, and the number of civilizations it may hold.

This book is my attempt to assemble 16 of the really big ideas in astronomy, planetary science, and cosmology that have exploded in terms of understanding over the past generation.

I hope that you enjoy it and know that we are still early in the game, despite these huge recent advances, in really knowing the vastly huge universe that surrounds us.

Carl Sagan used to lament that, "ninety-nine percent of people on Earth are born, live their lives, and die without realizing their place in the cosmos." I hope that this book helps to decrease that percentage just a little, and bring about the majestic wonder of the universe to more people on our little blue planet.

Acknowledgments

This book owes itself to the help, encouragement, donation of wise guidance, and patient endurance of many wonderful people. First and foremost is the time and support given by my exceptional wife Lynda Eicher, the outstanding encouragement and humor over the project from my son Chris Eicher, and the wonderful intellectual discussions and philosophical inspirations about astronomy, cosmology, and the meaning of it all that take place every few days with my father, John Eicher.

This book was originally inspired by Carl Sagan, many years ago, a friend and supporter whom we lost way too soon.

I also thank several friends, relatives, and associates who encouraged me on various aspects of this book or helped with related projects: Michael Bakich, Nancy Eicher, John Einberger, Garik Israelian, Thomas Kraupe, Brian May, Dan Murray, Martin Ratcliffe, Robin Rees, Grigorij Richters, Glenn Smith, Karri Stock, and Rich Talcott. I also thank my editor at Cambridge University Press, Vince Higgs, Rachel Ewen, Cambridge's publicist, and Karyn Bailey and Beata Mako of Cambridge University Press who have done such great work. Thanks to Kevin Keefe, Publisher of *Astronomy* magazine, for granting permission to use illustrations created originally for the magazine.

An enormous debt of thanks goes to a friend and hero in the field of astrophysics, Alex Filippenko of the University of California, Berkeley, and his lovely wife Noelle too, for their encouragement and support, and for Alex's gracious acceptance of the offer to write a foreword for this volume.

This project was also fortunate enough to receive some wonderful support from a stellar cast of astrophysicists, cosmologists, and planetary scientists who volunteered to read over the chapters for scientific balance and accuracy. That said, any resulting errors or discrepancies are solely my fault. But I am deeply indebted to my friends and colleagues:

Bruce Balick, *University of Washington, Seattle*
Robert A. Benjamin, *University of Wisconsin, Whitewater*
Alex Filippenko, *University of California, Berkeley*

Debra Fischer, *Yale University*
John S. Gallagher III, *University of Wisconsin, Madison*
James W. Head III, *Brown University*
Dan Hooper, *Fermilab*
John Kormendy, *University of Texas, Austin*
Abraham Loeb, *Harvard University*
Alfred McEwen, *Lunar and Planetary Laboratory, University of Arizona*
Rocky Kolb, *University of Chicago*
Michael R. Rampino, *New York University*
Martin J. Rees, *Institute of Astronomy, University of Cambridge*
Adam Riess, *Johns Hopkins University*
Seth Shostak, *SETI Institute*
Paul D. Spudis, *Lunar and Planetary Institute*
Alan Stern, *Southwest Research Institute*

I also owe a special debt of gratitude and thanks to another colleague of mine, *Astronomy* magazine's newest editor, Eric Betz, who joined our publication in the summer of 2014. Eric graciously agreed to take on the big task of finding illustrations for this book, arranging them into a wonderful sequence, and writing captions for the project. Eric's talented visual eye has helped to make this book into what I hope, for a variety of reasons, will offer a pleasurable way to spend some time in a comfortable chair, intellectually exploring the vast cosmos around us.

Chapter 1
The awakening of astronomy

Some 7 million years ago, a group of creatures made its way across the plains of central Africa. Resembling at first a collection of savannah baboons, the 30 or so beings shuffled along as dusk began to fall over a clearing in what we now call Chad. Adult females and flocks of offspring made up the nucleus of this foray, with a few mature males following up and looking for mating opportunities. As darkness began to fall, the group approached a cave that held a common shelter, and light from the Moon blazed down onto what now appeared like black, slumped forms – dirty, disheveled, hairy, and marked by spots of blood from the day's successful hunt.

These earliest hominids, perhaps *Sahelanthropus*, were the first bipedals, and walked more or less upright. They stand among the earliest creatures known from around the time of the human/chimpanzee divergence, when our ancestors began to make their own lineage that would one day lead to *Homo sapiens*. As these creatures, primitive by today's standards, shambled back to their nightly caves, they no doubt occasionally looked skyward, at the Moon and the stars. Perhaps they wondered what those lights in the sky meant. Somewhere around this time, some kind of creatures like *Sahelanthropus* became the first early human ancestors to ponder what space above meant to them.

Human knowledge about astronomy awakened painfully slowly, however. The earliest thoughts about the sky resulting in evidence we can examine were probably related to calendars and monuments, or tools for the planting and harvesting of crops, once humans became farmers. Although they weren't observatories per se, Stonehenge and other ritual Neolithic and Bronze Age sites betray a basic knowledge of the heavens. Egyptian, Spanish, Mexican, Irish, and Scottish stone structures nicely record celestial alignments. Stars no doubt also served as navigational tools for early explorers on land and on water.

Like all sciences, astronomy emerged from a primitive root that stunted progress for centuries – in this case, astrology. But as ideas emerged slowly and the astronomy of Antiquity began to inch forward, astronomy was a science of classification. For centuries, the idea was simply to look at things

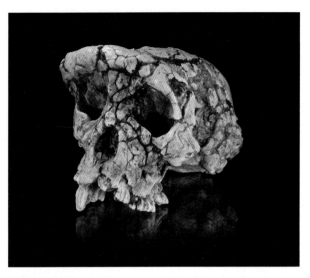

Figure 1.1 This cast from a partial skull is one of the few known *Sahelanthropus tchadensis* specimens.
Didier Descouens

and try to begin to understand them by sorting and noting similarities and differences. The same process governed other sciences too, as with studying seashells, mineral specimens, the skeletons of cats and dogs, or a thousand other things. But astronomy offered one unique difference – the sky was open to all comers, amateur and professional alike. Unlike virtually all other sciences, astronomy shared the same laboratory with everyone.

At first, mind you, the classification process was extremely basic. For centuries, from late Antiquity to the seventeenth century, astronomers really were obsessed with the motions of the planets. Were the movements of the planets regular and predictable and graceful? Could future positions of the planets be predicted? Astronomy was centered on this issue.

The first great leap in observational astronomy came on an autumn night in 1609, when Italian physicist, mathematician, and astronomer Galileo Galilei (1564–1642) climbed to the top of his house to make a rooftop observation. A few weeks earlier Galileo, ever the ambitious teacher and inventor, had heard troubling news. Dutch opticians had made a device using curved lenses that could magnify distant objects, making them seem closer. Galileo heard this while he was in Venice, and believed that cheap so-called telescopes were even showing up for sale on the streets of Paris.

The problem was that for some considerable time, Galileo had himself envisaged creating such a device. And he had good reason to produce such an instrument as for some time he had intended to impress Leonardo Donato (1536–1612), the Doge of Venice. After he heard about the telescope's invention, Galileo rushed home to his Padua workshop, and over the course of little more than a day, created his own 3x telescope with a lens about 1-inch in diameter, simply from what he had heard. He demonstrated the instrument to Venetian officials on August 25, 1609, climbing the Campanile on St. Mark's Square and showing his guests naval ships on the horizon. The military value of the telescope was immediately apparent. Galileo instantly became a star.

When he climbed to the top of his Padua house a few weeks later, Galileo initially looked at the spires of the Basilica of Saint Anthony of Padua, the massive church near his house. Then, in a fateful moment in the history of science, he swung his telescope's field of view over to the Moon, which lay nearby. In doing so, Galileo made the most influential early telescopic observation of a celestial body. (Yes, Englishman Thomas Harriot [ca. 1560–1621] apparently sketched the Moon with a telescope some 4 months earlier, but Galileo's observations were the ones with a towering, lasting effect on the history of science.)

In an instant, Galileo saw the Moon as a pockmarked, imperfect disk, with dark "seas," craters, and mountain ridges. By the end of 1609, Galileo had a simple telescope with a magnification of 20x, and made his legendary observations of the four "Galilean" moons of Jupiter, which he discovered; of sunspots moving across the solar surface; of the stellar structure of the Milky Way; and of the phases

Figure 1.2a Galileo Galilei presents his telescope to Leonardo Donato in this 1754 painting by H. J. Detouche.
H. J. Detouche

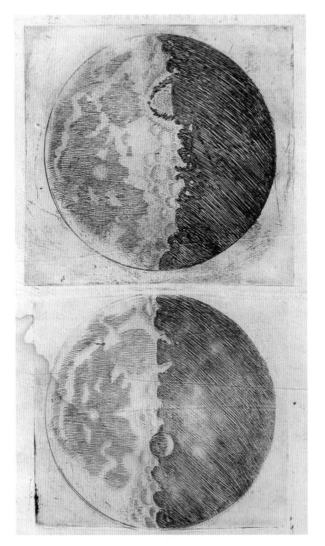

Figure 1.2b This early Galileo sketch shows the lunar surface as seen through an early refracting telescope.
History of Science Collections, University of Oklahoma Libraries

of Venus, which confirmed the Sun-centered view of the cosmos that Polish astronomer Nicolaus Copernicus (1473–1543) had proposed 66 years earlier.

Galileo sparked a revolution in that, from that moment forward, empirical observations, the creation of careful hypotheses, experimentation, and repeated observations would rule the day. What would become the systematic scientific method has its roots in Galileo's early telescopic observations, and the science of astronomy gained momentum. German astronomer, mathematician, and astrologer Johannes Kepler (1571–1630) also transformed the standard view with his three laws of planetary motion, which changed the game from simply predicting geometrically the movements of the planets to trying to understand the mathematical relationships behind the movements.

When English physicist and mathematician Isaac Newton (1642–1727) came along, the basis of astronomy in rational physics solidified. Newton's *Mathematical Principles of Natural Philosophy* (Latin: *Philiosophiae Naturalis Principia Mathematica*, or *Principia* for short), published in 1687, laid down classical mechanics, the laws of motion, and the law of universal gravitation – all keys to understanding the universe in a fundamental way.

The eighteenth and early nineteenth centuries were a time of discovery, of cataloging stars, comets, asteroids, and deep-sky objects. German–English musician and astronomer William Herschel (1738–1822) discovered the planet Uranus from his garden in Bath, England, in 1781. His son John Herschel (1792–1871) discovered and cataloged numerous objects in the Southern Hemisphere sky. In 1846, German astronomer Johann Gottfried Galle (1812–1910) discovered the planet Neptune, based on mathematical calculations by the French mathematician Urbain Le Verrier (1811–1877).

And then came another revolution. In the second half of the nineteenth century, scientists started attaching prisms to the eyepieces of telescopes, and the era of astrophysics was born. Now, a rainbow-like spectrum would betray the chemical composition of stars, planets, the Sun, nebulae, and other objects. Now astronomers had to add the expertise of being a laboratory physicist to their resumes as well as all the other previous skills. And the explosion of information that would come from spectroscopy and also photography would send the knowledge base of the universe skyrocketing.

Larger telescopes were in the offing, too. Before the era of World War I, the biggest telescope on the planet was the 72-inch "Leviathan of Parsonstown," a speculum-metal mirror reflector constructed by William Parsons, 3rd Earl of Rosse (1800–1867) in the 1840s, at Birr Castle, County Offaly, Ireland. By the turn of the twentieth century, American astronomer George Ellery Hale (1868–1938) was becoming a driving force behind several large telescope projects, resulting in the construction of the 100-inch Hooker Telescope at Mount Wilson near Los Angeles, California, and ultimately the 200-inch Hale Telescope at Palomar Mountain, California, following Hale's death.

Large telescopes coupled with the tools of astrophysics resulted in quantum leaps forward in human knowledge about the cosmos. A "great debate" occurred in 1920 between American astronomers Harlow Shapley (1885–1972) and Heber D. Curtis (1872–1942) over the nature of "spiral nebulae" and the distance scale of the universe. Three years later American astronomer Edwin P. Hubble (1889–1953) discovered a Cepheid variable star (of known intrinsic brightness) in the big spiral nebula in Andromeda, and discovered it must be much larger and more distant than anyone had imagined. It became the Andromeda Galaxy (M31), and the first step in understanding the nature of galaxies and the distance scale of the universe had been taken.

Figure 1.3 Completed in 1948, the Hale Telescope at California's Palomar Observatory
towered as the largest telescope in the world for more than 25 years, with a mirror twice the
diameter of the next largest instrument. It remains active in research today.
Caltech/Palomar Observatory

Seven years later, a Kansas farm boy who had taken up as an astronomer at
Lowell Observatory in Flagstaff, Arizona, Clyde W. Tombaugh (1906–1997),
discovered a distant body that was identified as a ninth planet in the solar
system and named Pluto. Although Pluto was to be demoted to dwarf planet
status by the International Astronomical Union in 2006, the first step in understanding the complexity and numbers of icy bodies in the solar system was
complete.

Meanwhile, in the midst of these pure discoveries, others were revolution-
izing astronomy simply through brilliant thoughts. Just after the turn of the
twentieth century, in Bern, Switzerland, aspiring German physicist Albert
Einstein (1879–1955) conducted thought experiments during his mostly mun-
dane life as a patent clerk. What if the streetcars he rode every day were
moving as fast as the speed of light? What if the elevator he rode in dropped
at the speed of light? What would the implications be? Einstein's mental
explorations ultimately led to the Special Theory of Relativity (1905) and the
General Theory of Relativity (1915), which rewrote our understanding of the
cosmos.

And yet, with all these breakthroughs, fundamental mysteries abounded in
the human understanding of the universe around us, through most of the rest of
the twentieth century.

But yet another revolution has taken place, one that is far more sudden and remarkable. Over just the past decade, astronomers, planetary scientists, and cosmologists have answered or are closing in on answers to some of the biggest questions about the universe.

This revolution is rapidly recasting what we know about the universe around us. Given this flood of new findings and new ways to understand nature, we can now say that we live in a place redefined by our fresh knowledge – a new cosmos.

These redefinitions address questions that are as old as time and as fundamental as space itself. They answer timeless mysteries about the universe's origin, its fate, its size, its shape, and its age. We now know about the relative numbers of planets around us in nearby parts of the Milky Way Galaxy. We know about the barred spiral structure of the Milky Way and about its future collision and merger with our neighbor in Andromeda. We know about the ubiquitous nature of black holes in galaxies. We know about the cosmic distance scale of our galaxy, about the vast stretches of empty space that surround us inside the Milky Way and far beyond it.

The astronomical revolution of the early twenty-first century casts a wide net. One major future event we now understand pretty clearly is how the Sun will die. The solar system is about halfway through its normal existence, some 4.6 billion years old. The Sun is, of course, a nuclear fusion reactor, and when it runs out of elements to fuse into heavier elements, it will become a red giant star, swelling outward and engulfing the inner planets, some 6 billion years from now. Following the red giant stage, the Sun will transform into a planetary nebula, a cocoon of glowing gas surrounding the dead Sun, which will then be a planet-sized lump of carbon and oxygen containing about 50 percent of the Sun's original mass.

At this stage, the Sun will be furiously bright. The last bits of helium burning within the star will fling the star's outer layers off into surrounding space, forming a so-called planetary nebula. Eighteenth-century observers named these disk-like glowing spheres planetary nebulae, believing they looked like planets. As a planetary nebula continues to form, episodic bursts of spasmodic burning eject more shells of gas away from the star, some at higher velocities than others, and the photoionization between these "burps" of gas causes them to glow like a fluorescent light bulb.

This produces the planetary nebula we can see in the sky (and in modern telescopes), each of which lasts for some 50,000 years before dissipating into the interstellar medium. Planetary nebulae serve as a recycling mechanism for turning the gas from many ordinary stars forward toward future generations of new stars, when it eventually compresses into a star-forming molecular cloud, pulled by gravity's inescapable force, and nuclear reactions begin, creating a newly born star.

Just as we can forecast the distant future of the Sun, we can also predict what will eventually happen to life on Earth. Whenever I give a talk to an astronomy group, I like to ask about this next question. Because the Sun is about halfway through its lifetime, simple logic suggests that life on Earth should be about halfway through its existence too. That seems a reasonable assumption, even among highly informed astronomers and astronomy enthusiasts. But such is not at all the case.

Ever since the 1990s, astronomers and Earth scientists have been analyzing the question of the long-term habitability of the terrestrial planets. The question of why Venus is too hot, Mars is too cold, and Earth is just right for life is the driving factor in understanding climates on these worlds. Strangely, early Earth seemed a perfect place for life to take a foothold, despite the so-called faint early Sun paradox. That is, some 3 to 4 billion years ago, the Sun produced only 60 or 70 percent of the total radiation it does today, yet life got going on early Earth, despite the faint environment that included liquid water and abundant carbon dioxide.

In fact, some of the earliest life known on Earth, 3.4 billion years old, comes from the Strelley Pool rock formation in Western Australia, where researchers discovered microfossils in 2007 and made their analyses public in 2011. These primitive bacteria fed on sulfur and were discovered in sandstones that, several billion years ago, formed a shallow water beach or estuary. Some researchers believe that other primitive microbes in rocks at Isua in Greenland show the imprint of microbial life dating to 3.75 billion years ago.

What we now know for certain is that the Sun is a variable star, and that its overall radiation output is steadily increasing over time. Recent work shows that in a far shorter timespan than had been previously imagined – perhaps a billion years or less – the Sun's radiation will increase to the point of boiling the oceans off planet Earth. At that point, it will mark the endgame for life on Earth. Given the knowledge of life on Earth existing for at least 3.4 billion years already, we can say the story of life on Earth is perhaps already 80 percent written. We are already in the late chapters of life's adventure on our planet.

Only in the last decade have planetary scientists really come to grips with the formation of the Moon. Years ago, astronomers struggled with this idea, in part because the Moon is so large as satellites go compared with Earth itself. They proposed "co-accretion," in which Earth and the Moon formed independently and then came together gravitationally: "capture," in which Earth gravitationally dragged the Moon into orbit after its formation and a near-miss encounter; and "fission," in which Earth's interior belched out the Moon like the splitting of a cell. None of these ideas fully convinced astronomers or matched what planetary scientists were observing with the Earth–Moon system.

Compelling evidence about the Moon's origin came from analyzing Moon rocks returned to Earth by the Apollo astronauts. Tests on oxygen isotopes locked inside tiny crystals in the rocks startled planetary scientists at first because the isotopes were identical with many Earth rocks. Scientists also believe the Moon had a very hot birth. At first perplexingly, the more scientists found out about the Moon rocks, the more the rocks began to resemble rocks from Earth's mantle, the outer shell of rock on our planet.

Over the 1980s and 1990s, lines of evidence from the Apollo samples began pointing toward a radical conclusion. Called the Giant Impact Hypothesis, the accepted story of the Moon's formation suggests that 4.6 billion years ago, two planets floated in the space now occupied by the Earth–Moon system. Proto-Earth had about 90 percent of its current size and mass, and a Mars-size planet also existed, one that astronomers now call Theia (in Greek mythology, mother of the Moon goddess Selene). Planetary scientists believe some 4.53 billion years ago Theia struck Earth, creating a short-lived ring of debris that accreted into the Moon. The majority of Theia's mass accreted into Earth's mantle. Where did Theia go? You're standing on it.

It might be fair to say that planetary scientists are obsessed with Mars. Some 60 percent of the world's planetary exploration budget is devoted to the Red Planet, and for good reason – in following the water, we are tracing the evidence that may lead to the discovery of present or past microbial life, which would be a momentous discovery. At a minimum, we're bound to understand a great deal more about life in the universe, as everything scientists think they know about life includes the need for water (or another solvent) in order to make it work.

From multiple spacecraft missions, we know that Mars has had abundant liquid surface water in the past. The Noachian period on Mars, roughly coinciding with the Late Heavy Bombardment 4.1 to 3.8 billion years ago, a period of intense impact cratering on the inner planets, seems to have been a warm, wet period on Mars. Significant erosion and dissection by valleys on Mars point to this conclusion, along with the existence of long-defunct lakes and oceans marked by marine sediments. Large bodies of water must have occupied such martian areas as Hellas, Argyre, and the northern plains.

Certainly, Mars is now a cold, dry planet. So how did Mars go from wet to cold and dry, and are there important lessons on the Red Planet for the residents of planet Earth? The mechanism by which Mars warmed is not yet entirely clear. It seems that substantial warming by a carbon dioxide–water greenhouse gas cycle would not work if the Sun were as faint as it appears to have been in the early solar system. But perhaps the Sun was more energetic early on than planetary scientists believe. Or maybe other greenhouse gases contributed to early martian warming. Or maybe warm periods on Mars were episodic and local and/or regional, rather than planet-wide, over sustained periods. Certainly, Mars had a much thicker atmosphere in those times, or liquid water would not have existed on the surface.

Another strange mystery concerns our so-called "sister planet," Venus, which could hardly be any more different than Earth. Venus and Earth are about the same size and Venus has a complex weather system, but beyond those similarities, Venus is a hellish world beset by incredibly high temperatures of around 480 °C (900 °F), hot enough to melt lead. Moreover, the air

Figure 1.4 This Hubble Space Telescope view of the Red Planet, taken in 2007, is among the best captured from Earth. Although Mars has been long shrouded in mystery and science fiction, spacecraft have revolutionized public perceptions of the Red Planet since Mariner 4 completed the first flyby 50 years ago.
NASA, ESA, the Hubble Heritage Team (STScI/AURA), J. Bell (Cornell University), and M. Wolff (Space Science Institute)

pressure is bone-crushingly heavy, more than 90 times greater than Earth's. The atmosphere exists under so much pressure that, near the surface, it behaves like a transparent liquid with the fluidity of molasses.

Venus was first explored extensively, and mapped in nice detail, by the Magellan spacecraft in the early 1990s. Immediately, this mapping mission revealed some pretty amazing results that are still being worked on and puzzled over. Smooth plains created by major volcanic eruptions extensively cover the surface of Venus. The presence of sulfur in the venusian atmosphere suggests that occasional volcanic activity still takes place. But as the Moon and Mercury show us so well, the inner solar system has been heavily battered by small impacts – during the Late Heavy Bombardment and also, with less frequency, more recently.

The shock from Venus is that it has very few impact craters. The extensive surface flows of lava suggest the planet's surface is very young – perhaps 300 to 500 million years old. This is a planet that geologically turned itself inside out; the planet was nearly entirely resurfaced. Why? What caused such a radical, planet-wide event that changed the character of the entire world?

It has now been more than 80 years since Clyde Tombaugh discovered Pluto. I still find it amazing to ask an audience whether they believe Pluto is a planet or not. The feelings are very strong, mostly in support of keeping Pluto a planet, and mostly from older amateur astronomers – not the schoolchildren who were originally so offended by Pluto's 2006 demotion to dwarf planet. These are middle-aged men who are still steaming mad over Pluto's demotion!

But the issue opens the door to all sorts of important scientific questions of the moment. Pluto was the first object discovered that would ultimately belong to the Kuiper Belt, the region of small, icy solar system bodies that extend from the orbit of Neptune (about 30 astronomical units from the Sun) to about 50 AU from the Sun. The Kuiper Belt is named after the Dutch–American astronomer Gerard P. Kuiper (1905–1973), and was recognized in 1992 when astronomers found the first member (before Pluto's membership was recognized) (15760) $1992QB_1$.

Over the past 20 years, planetary scientists have discovered more than 1,000 Kuiper Belt objects. They believe perhaps as many as 100,000 such objects greater than 100 kilometers across inhabit the Kuiper Belt. Who would want the schoolchildren of the world having to memorize thousands of names of new planets? In terms of planetary astronomy, Pluto was classified to be with its brethren in the Kuiper Belt. These are icy bodies, comets, or ice-rich asteroids, along with a few dwarf planets, and they have certainly over the past few years rewritten the definition of the relationship between comets and asteroids, blurring the distinction of what once was a clean separation.

It is impossible to have been interested in astronomy in the last decade without hearing an almost constant stream of information about newly detected planets outside our solar system.

The business of detecting exoplanets is big business – several large groups of researchers are working systematically on searches for more and more worlds, and getting closer to finding planets analogous to Earth. As of this writing, 1,932 exoplanets in 1,222 planetary systems have been discovered, and another several thousand candidates are known from the Kepler Space Telescope, whose primary mission ended in mid-2013, but was restarted in limited fashion as "K2" a year later.

Astronomers detected the first planet-like object outside our solar system, a planet orbiting the pulsar PSR B1257+12, in 1992. Three years later scientists detected the first planet orbiting a main sequence star other than the Sun, 51 Pegasi, at a distance of 51 light-years from Earth. Consider that Kepler observed one patch of sky $10°$ on a side on the boundary of the constellations Cygnus and Lyra, and operated for just 4 years, looking principally at stars closer than 100 light-years. Most exoplanets discovered before Kepler were giant planets like Neptune or Jupiter. Kepler's candidates include many Neptune and Jupiter-sized worlds, but also "super Earths" (21 percent of those found, from 1.25 to 2 Earth radii) and Earth-sized worlds (43 percent of those found, less than 1.25 Earth radii).

Kepler looked for planets in one small area of sky, and not very deep into the galaxy. Extrapolating those numbers over the entire sky and throughout the whole disk of the Milky Way brings visions of enormous numbers of planets into view. In 2013, astronomers at the Harvard-Smithsonian Center for Astrophysics suggested "at least 17 billion" Earth-sized exoplanets could exist within the Milky Way. The number of potential worlds out there is staggering.

Planets in our galaxy are one thing – the structure of the galaxy itself is another. When we go outside and look at the glowing band of light that traverses the sky, we can appreciate the awe and wonder the ancients had for the Milky Way (*via Lactea* in Latin). But on Earth, we are hamstrung in envisioning the Milky Way and its structure from the outside by the simple fact that we live within it.

Ever since Hubble's discovery of a Cepheid variable star in M31 in 1923, astronomers have realized the basic nature of galaxies. They are vast collections of stars, gas, and dust, mostly at very large distances, and the universe contains staggering numbers of them. Deciphering the shape and design of our own Milky Way has proved a challenge. For decades, astronomers believed the Milky Way was a garden-variety spiral galaxy because they believed it had a flat disk populated by numerous hot, young blue stars. Those attributes seemed to match spirals viewed elsewhere in the cosmos.

An important step in mapping the Milky Way was confirmed in 2005 when astronomers from the University of Wisconsin used the Spitzer Space Telescope to conduct the GLIMPSE survey, short for Galactic Legacy Infrared Mid-Plane Survey Extraordinaire. For some years beforehand, the evidence for a central bar in the Milky Way had mounted. But the GLIMPSE survey imaged and catalogued 30 million sources to accurately map the galaxy for the first time. Further observations in 2008 revealed the Milky Way is a barred spiral galaxy with a strong central bar, two prominent spiral arms, and several smaller spurs. Only in the past few years have we known the true structure of our home galaxy.

The same year, 2008, brought the most detailed analysis of our galaxy's fate. For some years beforehand, astronomers had been studying the dynamics of galaxies in our Local Group, which includes the Milky Way and the nearest spiral to us, the Andromeda Galaxy. In 2008, American astronomer Abraham Loeb (1962–) and his collaborators produced a detailed study. Although on large scales, the expansion of the universe is moving bodies apart, on more localized scales, objects often move toward each other due to gravity and localized motions. For a long time, astronomers have known the radial velocity of the Andromeda Galaxy, its motion relative to us is about –120 kilometers per second. That is, it's moving closer to us by about the distance between Milwaukee and Chicago every second.

The Andromeda Galaxy is 2.5 million light-years away – about 24 million trillion kilometers – so nothing drastic will take place for quite some time. But Loeb and his collaborators studied and modeled the eventual collision that will occur between the Milky Way and the Andromeda Galaxy, some 4 to 5 billion years from now, which will result in a tangled mess of a giant galaxy they have dubbed "Milkomeda." The fate of our own galaxy is now clear.

As we understand the fate of our galaxy, of our Sun, and of life on Earth, cosmologists have looked to the universe's beginning to understand how we got where we are. In 2013, astronomers using the European Planck spacecraft (launched in 2009) unveiled a new round of cosmological findings relating to the satellite's all-sky mapping of the cosmic microwave background, the "echo" of the Big Bang. Planck's results follow, support, and deepen the precision of earlier findings from the Cosmic Background Explorer (COBE) satellite of the 1990s and the Wilkinson Microwave Anisotropy Probe (WMAP) of the 2000s.

Planck scientists announced they had refined the age of the universe at 13.798 ± 0.037 billion years, slightly older than was previously thought. So all the textbooks, astronomical papers, and popular stories about the cosmos will now have to use 13.8 billion years as the universe's age rather than the long-familiar 13.7. Importantly, the satellite's numerous findings also reflected on the makeup of the cosmos, in terms of baryonic ("normal") matter, dark matter, and dark energy. (More on these later.)

Over the past few years, astronomers have also clarified their understanding of the size of the universe. Given the age of the universe, 13.8 billion years, the fact that the speed of light is constant at 299,792 kilometers per second (186,282 miles per second), and a variety of other factors, it's possible to estimate the overall size of the cosmos. However, there is a proviso. If we accept Inflation Theory as most cosmologists do – the idea originated by American physicist Alan Guth (1947–) and Russian–American physicist Andrei Linde (1948–) in 1980 – that the very early universe hyperinflated in size, then the observable universe we can see is not close to the entire universe. (And the universe is larger than that which we see independent of inflation.)

Nonetheless, the best calculations indicate the comoving distance between Earth and the "edge" of the observable universe is about 46 billion light-years in any direction, and that the universe's diameter is at least 93 billion light-years. How can the universe be that large if the speed of light is fixed and the age of the cosmos is 13.8 billion years? Because the universe isn't a big empty box with things simply expanding inside it; rather, space itself expands interstitially over time. So that 1 centimeter that used to exist stretched into 2 centimeters, and so on.

And that brings us back to some of cosmology's mysteries. As early as 1932, Dutch astronomer Jan H. Oort (1900–1992) postulated the existence of an unseen matter to help explain the orbital velocities of stars in the Milky Way. A year later Swiss astronomer Fritz Zwicky (1898–1974) proposed the same idea to explain the "missing mass" in galaxies orbiting each other in clusters. Soon, astronomers realized a form of so-called dark matter must exist in the universe.

American astronomer Vera Rubin (1928–) pushed forward the study of dark matter when she examined the rotational velocities of galaxies in the 1960s and 1970s. Astronomers still don't know what dark matter is, but recent cosmological studies from the Planck satellite and other sources suggest 26.8 percent of the mass-energy of the universe is in the form of dark matter.

If the challenge of dark matter wasn't enough, astronomers observing distant supernovae upset the cosmological apple cart completely in 1998. In that year, two competing teams of researchers, the High-z Supernova Search Team and the Supernova Cosmology Project, each observed distant Type Ia supernovae. These very distant exploding stars shine with a known intrinsic brightness. The observations of these unusual stars produced a stunning result. They showed the expansion of the universe is itself accelerating over time.

This amazing discovery, which led to the 2001 Nobel Prize in physics and reset the central focus of cosmology, led to the recognition of dark energy – the term given to the still-mysterious entity that is causing the universal expansion to accelerate. Astronomers have no idea what dark energy is; only clues, and yet they know from Planck and other sources that dark energy makes up 68.3 percent of the mass-energy of the cosmos.

In astronomy, mysteries come and mysteries go. When I arrived at *Astronomy* magazine as a young editor in 1982, black holes were merely a rumor. This was despite the fact that a region of gravity so intense that not even light could escape had been theorized as early as 1784 by English naturalist John Michell (1724–1793). Yet evidence for black holes was exasperatingly slow in coming. Considered mathematical curiosities, black holes were described in a version of relativity produced by German physicist Karl Schwarzschild (1873–1916) in 1916. In 1958, American physicist David Finkelstein (1929–) described the event horizon of a black hole in mathematical terms.

Observational evidence for black holes was hard to come by. The first strong black hole candidate was x-ray source Cygnus X-1, discovered in 1972. But doubts remained about this star-sized black hole. It wasn't until research with big ground-based telescopes in the late 1980s and then the Hubble Space Telescope came along in the 1990s that evidence for black holes arrived. After Hubble observed the centers of active galaxies such as M87 in Virgo and many others, the evidence for black holes was concrete. Over the past few years, it has become clear that almost all normal galaxies, dwarfs aside, have a central, supermassive black hole. (In our Local Group, M33 does not have one.) For most galaxies, these black holes were very active early in the history of the galaxy's life, and as the black hole has lost material to "feed on," it has become quiescent, going to sleep in a manner of speaking. Such is the case with the Milky Way's own central black hole.

And that ultimate question has also come into focus over the past several years: What is the fate of the universe itself? When the vast majority of cosmologists came to accept the Big Bang theory as the correct cosmological model, in the 1960s, the question of how the universe will "end up" became a fashionable line of thinking. The answer depends on several factors, including the mass-energy properties of the cosmos, the density of the universe, and its rate of expansion.

The general feeling among cosmologists is that, despite the hot, dramatic start to the universe, it will have an anticlimactic ending. The universe may have begun with a bang, goes the saying, but will most likely end with

Figure 1.5 A jet of energetic particles blasts from a supermassive black hole at the center of a galaxy in this artist's illustration.
NASA/JPL-Caltech

something very different. It appears the universe will expand forever, and that 100 billion years from now the cosmos will be unimaginably large and with the lights in scattered galaxies slowly winking out, leaving dim, red dwarf stars numbly burning like scattered coals in a far-away fireplace.

And yet the answer is not definite. Strange things could happen, like polarity reversals of dark energy that could halt the cosmic expansion. But the odds appear to favor a distant, cold, lonely, end of eternal darkness.

And recent work on astronomy and cosmology also merits a few words about the cosmic distance scale. Human civilization is in the midst of an amazing and unprecedented flood of astronomical data. We have never seen an era as productive as this one in terms of addressing the "big questions" of the field. Yet the gulf between science and the perception of reality by much of the public seems, if anything, to be widening. Too few people anchor themselves in reality in our culture that seems to be centered on laying back and watching a stream of mostly nonsense on TV, in movies, and online.

So here's a dose of harsh reality: the universe is REALLY BIG. The cosmic distance scale is almost unbelievable by human standards. We are on average 384,400 kilometers from our nearest cosmic neighbor, the Moon. The Sun is normally 149.6 million kilometers away. The outer edge of the Oort Cloud, the shell of comets at the perimeter of our solar system, is more than 1 light-year away. That's about 10 trillion kilometers. The closest star system to us, the Alpha-Proxima Centauri system, is about 4.2 light-years distant.

There are huge lessons for us in the distance scale of the universe, the likelihood of abundant life in the cosmos, and the possibility of ever getting to know that life. Because, let's face it: that's the ultimate question. Are we alone in the cosmos? How abundant is life? Microbes? Civilizations?

If money were no object and we could build the highest technology space-craft now imaginable, and power it with say, ion propulsion, we could hypo-thetically travel to the nearest star system, Alpha-Proxima Centauri, in something like 75,000 years. This is where science and science fiction meet. The logistics of actually traveling from star system to star system are staggering. How do you put enough Twinkies in the pantry for a 75,000-year journey?

Yes, technology changes over time and we or other civilizations may have or eventually have technological capabilities far more advanced than our present ones. But the laws of physics are still going to be the laws of physics. In reality, moving anything with mass at high velocities is very difficult. Photons can zip through the room at the speed of light and strike your eye, enabling you to read this page, because they are massless. Try to move anything with any appre-ciable mass at speeds of even a tiny, tiny fraction of the speed of light and you will encounter real trouble.

And yet we have seen incredible hints of the possibilities for life in the universe. Everywhere we look at extreme environments on Earth – undersea vents, frozen blocks of ice – we find hints of the incredible tenacity and resili-ence of living beings. The evidence for the formation of life on Earth suggests it got going early on, and started or restarted under incredible duress during or just after the Late Heavy Bombardment. Looking at the planets surrounding us in our galaxy, we see numerous examples of worlds where life could exist, and are starting to find Earth-sized planets.

We know the numbers game pretty well. The Milky Way Galaxy holds at least 200 billion and perhaps as many as 400 billion stars (astronomers don't know exactly how many because dwarf stars, which are numerous, are faint and difficult to see over large distances). We know of at least 100 billion galaxies in the universe. Let's say, to a first approximation, some 10,000 billion billion star systems could exist in the universe – and recall that's just the visible universe. That's an incredibly large number of potential places for life to exist on planetary systems.

Could we be the only civilization in the universe? It seems terribly difficult to believe. And yet we now know of only one world where life exists. The utter immensity of the distance scale of the universe does suggest that – although we may know of other civilizations – the sci-fi dreams of standing beside beings from another world, shaking their hands in Central Park, is not likely now or in the future. As American astronomer Carl Sagan (1934–1996) once said, the universe may be teeming with life, but it may be akin to two people on Earth, one in Australia and one in central North America. The two might live their entire lives without ever knowing the other existed.

But let's not look too far afield, at least not yet. We know life exists here on Earth. And life exists here because our star, the Sun, made it so.

Chapter 2
How the Sun will die

Every day we wake, arise, get out of bed, and go off to work, to school, or to play. And we do so with the assistance of the Sun, which gives us the power, the energy, the heat, and the light to exist. Indeed, if we want to think of an ultimate enabler in our lives from a scientific standpoint, then the Sun would be it. Life on Earth, of course, never would have been possible without the existence of our star.

Let's consider the Sun for a moment. We often hear that the Sun is an "average" star, an unspectacular, middleweight among countless numbers that are larger, more energetic, and more dramatic, or smaller, quieter, and longer lived. But is the Sun really an average star? What does that really mean? Moreover, we know the Sun is about halfway through its normal lifetime. What will become of our star when it runs out of fuel to burn? When will this happen? Will the Sun die slowly and quietly as the fuel runs out? Or is there a more spectacular fate in store?

Stars, of course, are plentiful in our galaxy and in the universe. And in our area of the Milky Way, our galactic neighborhood, things are pretty typical relative to other regions where stars are found. In the Milky Way Galaxy alone, astronomers believe about 400 billion stars exist. No one knows exactly because the abundant dwarf stars, so huge in numbers, are hard to detect over long distances because they are faint. Being the nuclear fusion reactors that they are, stars transform hydrogen into helium, and they emit huge amounts of radiation as a byproduct. They are nature's nuclear engines. As such, the Sun is a pretty good one. The Sun fuses roughly 700 million tons of hydrogen every second. That's the equivalent mass of 1.3 million fully loaded Boeing 747–8s.

The Sun is a spectral type G2V star, meaning it is whitish and is a so-called main sequence star, on the normal path of its evolutionary life, about midway from start to finish. The "G2" stands for the Sun's color-temperature classification, and the "V" means it's a main-sequence star. The Sun's spectrum peaks at yellow/green wavelengths, but the overall distribution of colors and intensities gives rise to what we perceive as "white light." Indeed, sunlight is the definition of white light. About 7.5 percent of stars in the neighborhood of our Sun are

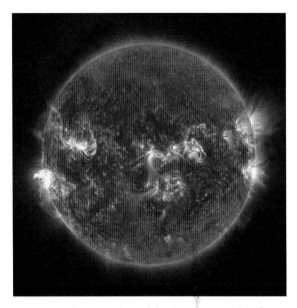

Figure 2.1 The active face of the Sun is covered in flares and prominences.
NASA/SDO

G-type main-sequence stars; Alpha Centauri A, Capella (a multiple star system –
its two brightest stars are sunlike) and Tau Ceti are among them. Stars can range
from about 0.05 to 50 times the mass of the Sun, although 93 percent are less
massive. Like rocks on a beach, most stars are small.

Although the Sun is small compared to the enormous scales of the Milky
Way, its gravity dominates the structure of our solar system, including the
size, speed, and period of Earth's orbit. The Sun's spherical mass spans about
1.39 million kilometers, meaning you could place 109 planet Earths side by side
across the Sun's face. Even Jupiter, the largest planet in our solar system, would
span only one-tenth of the Sun's diameter. The Sun's mass is approximately
2×10^{30} kilograms, meaning it holds about 330,000 times the mass of our planet.
Some three-fourths of the Sun's initial mass consisted of hydrogen, while
helium dominated the remainder, although significant amounts of oxygen,
carbon, neon, and iron also exist. This reflects the composition of the material
from which the solar system formed. This hydrogen is the fuel that supplies
the Sun's heat and light to Earth. It will have been converted to "waste" helium
in 5 billion years.

Thankfully, Earth formed in our solar system's habitable zone (or "Goldilocks zone"), allowing liquid water to exist on our planet and permitting the
eventuality of life. (Life could exist outside the habitable zone, but it would be
likelier within.) The Sun's mean distance from us is about 150 million kilometers
(= 1 Astronomical Unit), meaning light from the Sun's photosphere, its apparent
surface, takes 8 minutes 19 seconds to reach us. Once believed to be a relatively

insignificant star, the Sun is now known to be brighter than about 85 percent of the stars in the Milky Way, most of which are red dwarfs. The Sun, of course, dominates our sky, by far the brightest object in it, at visual magnitude –26.7. The Full Moon shines at a magnitude of –12.7, some 400,000 times fainter than the Sun. The Sun's diameter on our sky is about 32′, slightly more than half a degree, and it rises and sets with extraordinary regularity, giving us our annual calendar.

The basic physical characteristics of the Sun have been reasonably well known since the 1950s. Determining the age of the Sun, however, is a little trickier problem. Centuries ago, Isaac Newton began with thought experiments about how large masses of iron, which he supposed existed in Earth's core, would cool over time. In the nineteenth century, German physician and physicist Hermann von Helmholtz (1821–1894) and William Thomson, Lord Kelvin (1824–1907) applied thermodynamic arguments to finding the ages of both the Sun and Earth.

More recently, however, more definitive techniques for aging the solar system have come into play. Specifically, radiometric dating of rocks is a big one, since we believe the Sun and its planets formed at nearly the same time. This technique uses a comparison of observed abundances of radioactive isotopes within rocks or other objects, which have known rates of decay. They can thus be used to definitively date a wide range of materials, and anyone who questions that Earth, the Sun, or the solar system is as old as it is, for whatever motives they might possess, should be sentenced to radiometric dating classes at a conveniently close university.

Figure 2.2 A massive prominence erupts from the Sun, shooting a shock wave and particles across the solar system.
NASA/SDO/AIA/Goddard Space Flight Center

This radiometric technique is used to date very old rocks – more specifically, the oldest meteorites we have on Earth. Primitive carbonaceous chondrites from the outer solar system, analyzed on Earth and dated via radiometric techniques, yield ages of 4.567 billion years. They contain traces of stable nuclei of iron-60, an isotope created in supernovae, which suggests the gift of generations of supernovae that preceded the formation of the solar system.

The evidence suggests that perhaps a shock wave from the nearby supernova helped to create the formation of the Sun, and perhaps other stars with it, as the compression from the supernova shock pushed gas inward and gravity took over, pulling material in, eventually creating a spinning disk of gas and dust in which planets accreted and the solar system was up and running. When the hot, dense center of this juvenile disk accreted enough mass to ignite nuclear fusion, the Sun was born.

In a more direct way, astronomers have been able to estimate the age of the Sun itself. Several techniques have been employed by astronomers to do this. One is by studying computer models of stellar evolution, and understanding a wide array of properties of the Sun relative to countless other stars that have been studied, many of which are similar to the Sun, some of which are younger, and others older than our star. Further, solar physicists use a technique called nucleocosmochronology, which employs an examination of the star's heavy radioactive isotopes to affix a date to them, analogously to the radiometric dating of rocks.

And the age of the Sun can also be inferred through helioseismology, the study of wave oscillations through the Sun's interior, because the propagation of waves through the Sun depends on its composition, especially the ratios of hydrogen and helium in its core. So the amount of helium in the Sun's core is a measure of the star's age, based on simple calculations of how long fusion has been taking place. Each of these techniques converges on a value of 4.57 billion years for the age of our home star. Helioseismology also allows us to infer the inner structure of the Sun, much as the tonal spectrum of a large bell can be used to reconstruct its size, shape, and stiffness.

So our lives are beholden to a middleweight, whitish star about halfway through its normal lifetime, and we happen to be orbiting it at a distance of 150 million kilometers. Our beautiful lives on Earth can't go on forever, though, with this source of energy beaming its heat and light to us on and on and on. Nothing lasts forever. Not even the universe. What will happen to the Sun as it ages?

Consider for a moment one of the key tools in astrophysics, the so-called H-R diagram – or more formally, the Hertzsprung–Russell diagram – which was developed by and named after two astronomers around the year 1910. Danish chemist and astronomer Ejnar Hertzsprung (1873–1967) and American astronomer Henry Norris Russell (1877–1957) created the most famous scatter graph in astronomy, which plots stars' brightnesses on the Y axis versus their spectral types on the X axis. Thus, the groups of stars and wandering forms

Figure 2.3 Sirius A and its companion, the much smaller white dwarf Sirius B, are seen in this artist's illustration. The now tiny orb is the leftover remains of what was once a sunlike star. NASA, ESA, and G. Bacon (STScI)

revealed on this "map" allow us to learn a great deal about stellar evolution. The graceful patterns of stars on the diagram provide the vital clues against which we evaluate our understanding of stellar structure and evolution. Modern versions of the H-R diagram replace the spectral type of the stars with a precise color index measurement, and are termed color-magnitude diagrams (CMDs).

The diagram separates those incredibly large, powerful, and short-lived stars – giants and supergiants – in the upper right regions, and white dwarfs, burned-out embers of sunlike stars, forming a distinct and seemingly exclusive group near the bottom left. A long diagonal progression from top left to bottom right, however, shows normal main sequence stars, including the Sun just right of center. Some 95 percent of all stars are found on the main sequence, with the Sun near the middle and 95 percent of all stars beneath it. As it turns out, the main sequence is also a mass sequence, with rare, very massive stars at the top of commonplace, low-mass stars at the bottom.

When he created the H-R diagram, Henry Norris Russell believed the diagram depicted stars at different moments of their lives and that most stars began as hot, blue, and white, and progressed to cool, becoming increasingly red, along the main sequence track. While Russell's guess was very sensible, we

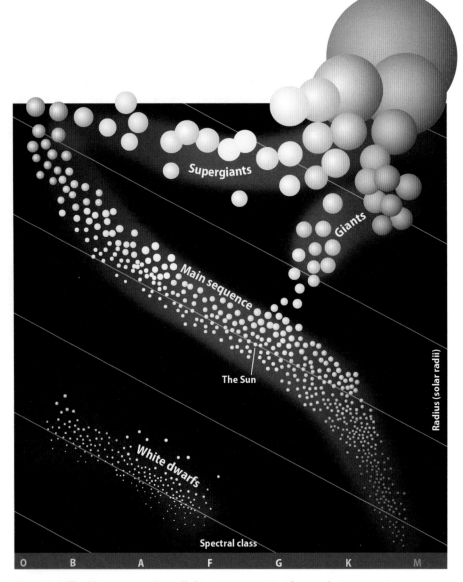

Figure 2.4 The Hertzsprung–Russell diagram was a significant advancement in understanding stellar evolution. The graph allowed astronomers to plot the main sequence of a star's life, where it spends most of its time fusing hydrogen in its core.
Astronomy: Roen Kelly

now know that the location of a star on the main sequence is determined solely by its mass while it fuses hydrogen to generate heat, not its age, location, or initial chemical composition. The main sequence is a sequence of initial mass.

For the Sun, we have a pretty good idea of what lies in its future.

As long as a star like the Sun burns hydrogen in its core, converting it into helium, its position on the main sequence doesn't change much. As it ages, it becomes slowly more luminous, and moves slightly to the upper right of the H-R diagram. Further, the star's mass is what really anchors it to a particular spot on the diagram, because its gravitational squeeze establishes the star's central temperature and rate of nuclear fusion. Because a star like the Sun has only so much hydrogen, it can remain on the main sequence for only so long, and then things get a little more interesting.

With a middleweight star such as the Sun, the rate of burning is moderate. Stars that are more massive have more hydrogen to fuse, but they also burn it at a much faster rate in order to sustain their huge luminosities, and so they "live their lives" more quickly – shine brightly, burn out fast, die young. Consider two imaginary cars of the same weight, an eight-cylinder Corvette and a Prius. The fast, showy Corvette squanders its fuel and quickly runs out of gas. The Prius runs further and for longer between fuel stops.

Most of the really massive stars that existed in the cosmos have already burned out. The massive stars that exist now are few and far between. Some are well known, however. The most massive star known, which has 265 times the mass of our Sun, is R136a1, a supermassive star at the heart of the Tarantula Nebula, a large star-forming region in the Large Magellanic Cloud, one of the Milky Way's satellite galaxies. This star is also the most luminous known, generating 8.7 million times the light of our Sun.

Stars like the Sun will shine for 10 to 20 billion years. But that's not to say they will be on the main sequence that long. The range of stellar lifetimes on the main sequence is really extreme. A spectral type O star, the hottest at around 44,500 K, and with a mass 60 times that of the Sun, lasts for less than 4 million years. (K = kelvin, the basic physical sciences measurement of temperature, with 1 K = –272 °C and –458 °F.) A B star might radiate at 15,000 K, have 6 times the Sun's mass, and live for 65 million years. A G-type star like the Sun exists on the main sequence for some 5 billion years. An M star, glowing at a mere 3,240 K, and with only 20 percent of the Sun's mass, lasts for trillions of years.

Nuclear fusion is the ultimate source of sunshine. And fusion exists only in a star's core. Away from the core, the pressure and temperature are insufficient to produce nuclear fusion, and the star's material is thinner and cooler. As with people, nothing lasts forever, and the life of a star must eventually come to an end. The fusion of hydrogen in any star is limited by its supply. In 1942, Brazilian physicist Mário Schoenberg (1914–1990) and the great Indian–American astrophysicist Subrahmanyan Chandrasekhar (1910–1995) demonstrated via modeling that a star cannot have more than 12 percent of its mass in the exhausted core. Therefore, they concluded, when a star depletes 12 percent of its hydrogen into helium, it exits the main sequence and begins a new kind of stellar age.

Ninety percent of stars are low-mass objects that will go on for a very long time, and many massive stars have already come and gone in our galaxy and in

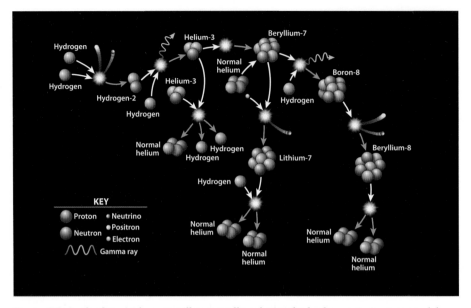

Figure 2.5 As the Sun evolves, it will eventually exhaust the hydrogen in its core used for fusion and move on to increasingly complex atoms, causing massive growth as the Sun swells into a red giant. When the Sun's fuel is exhausted, its core will collapse into a white dwarf. *Astronomy*: Roen Kelly

countless others. But the intermediate mass stars like the Sun offer an interesting forecast to our own solar system some 5 to 7 billion years from now.

What happens when a star like the Sun begins to run out of nuclear fuel? The star's engine turns off, almost like a fire burning out. The star's heat sustains its structure against gravity. Thus, when the nuclear fusion stops, the inner parts of the star resume the same collapse that occurred before fusion began. As this happens, the temperature in the core of the star rises. The star now begins to contract. Really massive stars undergo violent explosions like supernovae that catastrophically cast their contents far out into the regions surrounding them, seeding interstellar space with a rich broth of heavy elements. They briefly flash a dying message to the cosmos with the light of a billion suns. On the other hand, very low-mass stars like red dwarfs can burn a large proportion of their hydrogen and can last, like dim fireplace embers, for several trillion years.

Stars like the Sun face a more curious path toward demise. Amazingly, the average density of the Sun's material is pretty close to that of water. But the other extreme end states differ wildly. A red giant star, an enormously bloated envelope of gas, has a very dense core but an average density only a thousandth as great as water, whereas a white dwarf star, a highly compressed remnant, has a density nearly a billion times greater. And neutron stars and black holes, end states for very massive stars, have even higher densities.

As a dying star falls inward on itself, the sequence of coming events depends on its mass. The Sun and many other middleweight stars are destined to become red giant stars. The progenitors of these stars range from about a third of the Sun's mass to 8 times the Sun's mass. When one of these stars becomes a red giant, the outer envelope inflates and is shed outward, blown away by the radiation pressure of the evolving star. The star's diameter surges outward, its mass stays the same, its density drops, and its surface temperature plummets. At the same time, the core runs out of helium to fuse. So the core tries, and fails, to collapse. It can never fuse the ash of helium fusion, carbon, to generate more heat. The fate of the star's core is now set: it is a pure carbon hulk containing about half the star's original mass (60 percent for the Sun), packed into a volume no larger than Earth. It will later emerge as a cooling white dwarf star. During the red giant phase, the Sun will swell out to approximately the orbit of Earth. Of course, long before this point, life on Earth will have ceased, as we will see in another chapter.

The timing of the Sun's departure from the main sequence is something on the order of 5 billion years from now. It will continue through its red giant phase for some time, however, perhaps through 7 billion years from now, by which time it will reach its maximum diameter. During this process, it is possible that Mercury, Venus, and Earth will be destroyed, or at least heated to porcelain, although they may well migrate outward. At that point, the Sun will span at least 150 times its present size.

The dying star isn't completely dead yet, however. Gravity pushes shells of hydrogen and helium to densities where nuclear fusion can begin, and they are ignited, albeit very briefly. When this happens with the Sun, it will be some 30 percent brighter than it is now. At its greatest, the Sun will briefly be 2,100 times brighter than we see it.

Some 7 billion years into the future, the Sun will transform into a spectacular end state, creating a brief visual show in our part of the galaxy. At the end of its red giant phase, the Sun's core will really die. It will then transform into a white dwarf, a degenerate stellar remnant that is incredibly dense. This is the fate of nearly 97 percent of stars in the Milky Way – only the most massive 3 percent will become neutron stars or (more rarely) black holes.

A white dwarf has a maximum mass of about 1.4 solar masses and is initially hot but quickly loses its heat because it has no source of new energy. It simply sits in space and radiates away. The oldest known white dwarfs are still radiating at a few thousand kelvins, but are cooling all the time. The incredible part of the equation is that this white dwarf is no larger than a small planet, a tiny, burned-out shadow in the place of a formerly ferocious nuclear engine. At this point, the Sun will contain about 60 percent of its original mass and will be only slightly larger than Earth. The white dwarf Sun's collapse will have been halted by electron degeneracy, a principle of quantum physics that prevents subatomic particles from simultaneously occupying the same quantum state.

The Sun will put on a final show in the few thousand years before it succumbs to fading white dwarfdom. It will eject its outer layers as its nuclear

Figure 2.6 The Helix Nebula lies some 650 light-years away in the constellation Aquarius and consists of the remains of a sunlike star. As the star aged, it turned to helium for fuel before finally exhausting the heavier element and pushing its outer layers into space in a grand eruption.
NASA/JPL-Caltech

energy sources deep inside sputter. The fluffy outer layers that initially fill the volume between its Earth-sized core and its Earth's-orbit-sized surface will peel off, forming a planetary nebula much like a sheep shorn of its wool.

The planetary nebula will consist of a glowing series of shells of gas that expand outward at various speeds, putting on a celestial show for nearby parts of the galaxy. About 3,000 planetary nebulae exist right now in our galaxy, and a few hundred of the brightest are favorite targets for amateur astronomers armed with backyard telescopes. The Dumbbell Nebula (M27) in Vulpecula, the Ring Nebula (M57) in Lyra, the Helix Nebula (NGC 7293) in Aquarius, and the Owl Nebula (M97) in Ursa Major are some of the brightest.

It was the famous German–English astronomer William Herschel (1738–1822) who named these objects planetary nebulae. The discoverer of Uranus, Herschel, was intimately familiar with the appearance of the disks of planets in his telescopes, and believed these objects mimicked the appearances of planets perfectly – but they didn't move relative to the fixed stars. So in the 1780s, around the time of his discovery of Uranus, he coined the term, even though, of course, these objects have no relation to planets.

As the dying star lumbers on, the final chaos results in fits and starts of so-called helium burning, which casts the star's outer layers outward into interstellar space, producing the planetary nebula. This process isn't a clean, single stage shot, either. A planetary nebula forms from several episodes of "coughing spasms," at greater or lesser velocities, and this creates a large bubble of glowing gas, often in concentric rings, ranging outward into the galaxy starting at speeds of about 15,000 kilometers per second when the outermost layers are ejected and eventually reaching 1,000 kilometers per second as the atmosphere of the white dwarf itself comes off in a searing, low-density wind. That is, the star loses its huge, cool, dull-red outermost layers, eventually exposing the inner, tiny, blue-hot white dwarf to view in a cosmic instant, a mere few thousand years. As this happens, the ejected outer layers lumber outward, pushed by faster winds from the inside, to form the nebula. The whole affair is much like the sputtering of a gas-powered lawnmower that is gasping for fuel as its tank empties.

In rare cases, the last helium flash splatters dust and gas outward at higher velocities, and shapes planetary nebulae by setting up collisions between the new "bullets" of material and the older, more slowly expanding bubbles. For some preplanetary nebulae, this creates a bipolar shape as the fast-paced material shoots outward. It also creates dark dust lanes. The dust makes up a small fraction of the total amount of material, but it enriches the interstellar medium with carbon and silicate-rich grains and richly outfitted carbon-based molecules. The dust particles are tiny, only a thousandth of a millimeter across, but help to give the planetary nebula its appearance by reflecting light from the dying star, just as a reflection nebula does.

At this stage, we have a newly minted planetary nebula, sometimes called a preplanetary nebula, which is in the thousand-year process of brightening and becoming visible from a distance. As the nebula's gas continues expanding, making the nebula increasingly visible, the star is in the process of shedding its outer layers and heating to a temperature of 30,000 K. Formally, a planetary nebula is ionized gas that is produced when the exposed stellar surface surpasses 30,000 K.

The preplanetary nebula itself is also in a state of rapid evolution. Dimly entering its galactic surroundings, the nebula is now purely a reflection nebula, glowing dully from the scattering of starlight. As they scatter starlight, the dust particles also shield light from the infant nebula within, until the growing planetary nebula emerges like a chrysalis emerging from its cocoon, as stellar winds drive the obscuring dust outward.

Now, you might say, the fun begins and the planetary nebula gets its personality. When the star reaches 30,000 K, it begins to ionize the surrounding gas, heating it as ultraviolet photons strip electrons from neutral atoms, causing the gas to fluoresce much as a high-voltage current makes lightning visible as atoms in the air are stripped of their electrons. This brightens the nebula greatly and makes it visible from afar, and also aids in defining its shape as the gas and dust continue to interact.

But the ways in which the interactions take place are not yet perfectly understood. The models of how they should be taking place do not yet perfectly match the observations. Astronomers believe that radiation pressure from the central star should push dust outward in a radial fashion, spherically, and yet anyone who has looked at many of the spectacular images of planetaries made with the Hubble Space Telescope and other instruments can see that this is not always the case. Very few of the young planetary nebulae observed show dusty outflows that are round. What exactly is going on here?

Other peculiarities have also cropped up. Doppler shifts, measured for the outflows of carbon monoxide emissions in preplanetary nebulae, show that the outward momentum of the gas is something like a factor of 10,000 times faster than astronomers had predicted it to be, based on the conventional wisdom of how the material would be pushed away by radiation pressure. This is a mystery yet to be solved.

But it is clear that this so-called superwind is a precursor to other events in the life of a planetary nebula. Additionally, a so-called fast wind can blast off the dying star with a velocity of as much as 1,000 kilometers per second – a lower density, higher velocity final blast that helps to sculpt the planetary nebula's final shape from the inside.

This sets up some of the intricate architecture we see in telescopes – the fast wind barrels through the material ejected by the superwind, creating a shock front where the two collide and the appearance of a hollow cavity between the star and the leading edge of the most intensely glowing gas. The hollowness is partly an illusion – the bubble is filled with very tenuous and very hot gas that is heated to about 10 million K by the fast winds from the atmosphere of the white dwarf. Most of the atoms inside the bubble are completely stripped of their electrons. Some of the stars and some of the bubbles are observed in x-rays. The growth of the bubble is like a supersonically expanding balloon. Gas upstream from its perimeter is swept up into a dense, bright shell.

In the planetary nebulae they have studied with large telescopes, astronomers have detected substantial x-ray emissions inside the hollow shell, betraying the presence of extremely hot material heated by the winds to more than a million degrees. The expanding bubble of hot gas continues moving outward, and the shocked front – the main ring of bright material – pushes out to expand further, compressing more and more material that initially was outside the shell. When the faster winds reach the outer reaches of the shell of the old superwinds, the planetary nebula can "burst," blowing material into the surrounding interstellar medium like air from a ruptured balloon shooting out during the first moment it has been popped. This can produce strange lobes that reach far away from the central, bright nebula.

Ultimately, a planetary nebula can last in its proudly glowing state for perhaps 50,000 years before its material disperses into the interstellar medium. It "switches on" when the dying star heats up sufficiently to ionize the surrounding gas, and the fast wind breaks away the star's outer layers, exposing intensely hot, inner regions. After a few thousand years, the star's surface becomes noticeably bluer, and its temperature maxes out to nearly 100,000 K, many times hotter than the Sun.

The star's blazingly hot "surface" is an immense source of ultraviolet radiation, causing the nebula to switch on and become easily visible. Some half of this radiation it is dosed in gets converted into visible light, enabling the stellar death shrouds to be seen from halfway across the galaxy. Eventually, the nebula expands so much that it fades from view and its contents slowly merge into the interstellar medium.

Astronomers developed the first detailed models of planetary nebulae in the 1990s, but soon observations were showing a somewhat different story. Images produced with the Hubble Space Telescope revealed many variants on the basic possibilities, leaving scientists with the task of figuring out what was going on that was more complex in nature than the models predicted. Faced with difficulties matching the observations to models, astronomers continue to interpret what they are seeing in planetaries at a detailed level. The formation of asymmetrical axes, small knots, and other features challenge what we know about gas dynamics and the late stages of stellar evolution.

The relationship between preplanetary nebulae and full-blown planetary nebulae also offers some shape-related mysteries. Half of the preplanetary nebulae are shaped elliptically, and many feature a broad band of dust bisecting their central regions. With full-blown planetaries in mid-life, however, a mere 10 percent are shaped as bipolar objects, and there is very little central dust involved. What happens in the evolution of the nebula from a very early stage to change the shapes of so many planetary nebulae? Astronomers do not yet know.

Maybe part of what is happening with planetaries is that many are formed from double star systems. Some astronomers believe that as a red giant begins to come apart, a second star nearby pulls gravitationally on the loose outer layers of the aged giant, creating a bipolar shape. If this happens, astronomers reason that tidal forces pull the star's outer layers toward the companion during the helium flash, and much of the material overshoots the position of the companion, which has moved outward. This pushes material outward in the spiral pattern we observe on the sky. Some of the material might fall onto the companion, become heated, and form a disk, which could then create swirling patterns of material flowing outward. Or, perhaps the red giant star swallows a companion as it swells outward, which also creates some interesting possibilities for shaping a planetary nebula. The smaller star would be absorbed into the red giant, churning up the internal, loose, swelling portion of the larger star. This would also produce a disk that would help to shape the eventual surrounding nebula. It could also transfer the momentum from strong stellar winds coming off the red giant, producing fierce winds cascading off the star and helping to sculpt the planetary nebula we see. And if a smaller star spirals in like something going down a drain, then a huge amount of mass may create a spectacular, energetic burst of energy outrushing once it hits the core, perhaps explaining why the final outflow from some planetary nebulae is so strong and violent.

Magnetic fields can also help to shape planetary nebulae. The extraordinary heat and depth of material mixing in the helium flash could churn up intense magnetic fields as material flows out from the red giant's core, and these magnetic fields could help to guide the material flowing outward in specific ways. The computer modeling to understand how magnetic fields would behave is still in its early stages, and guessing how magnetic fields would stretch under the circumstances of dying stars and gaseous outflows is a risky business. But astronomers suggest that given what they know about magnetic fields, coupled with the shapes they are seeing in planetaries, the two are very likely to be importantly linked in many objects.

Some 230 years ago, William Herschel coined the term planetary nebula to describe what he saw in his telescope. In the 1990s, computer modeling and observations with many telescopes, chiefly the Hubble Space Telescope, produced a new paradigm in understanding these unusual, temporary, beautiful objects in the galaxy. We know that one day the Sun will become a planetary nebula, and only the future mixing bowl of our solar system's unique recipe will determine what shape that nebula takes on.

The last several years have focused astronomers' understanding of these mysterious objects more clearly. Known to be the temporary end result of the majority of stars in the galaxy, after which a white dwarf star will be left in isolation, planetaries do show us a mesmerizing array of shapes. Certainly,

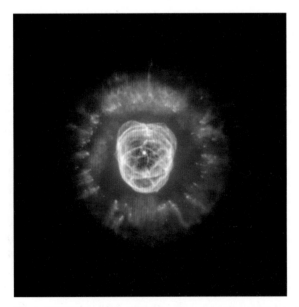

Figure 2.7 William Herschel discovered the bizarre light of the Eskimo Nebula (NGC 2392), now known to contain the remains of a dying sunlike star. A strong particle wind pushes outward from the central star, giving this planetary nebula the look of a person's head wrapped in a parka.
NASA/ESA/Andrew Fruchter/STScI/ERO(STScI/ST-ECF)

single stars produce a dazzling array of shapes simply from the details of how material moves away from the dying sun and interacts with its surroundings. Many astronomers certainly believe that many of the more complex and extreme planetary nebulae form from the interaction of a double star system, one star of which is an elderly red giant.

Although we don't know the specific details of how our solar system's planetary nebula will be shaped, we know it will be in the endgame for the Sun. It will be a magnificent, gently glowing cocoon of gas fluorescing for a few tens of thousands of years, signaling to the neighborhoods near us, in our Milky Way, that the star called the Sun has lived a normal life, but now is no more.

Chapter 3
The end of life on Earth

Our planet experienced a very violent early history. As planetesimals, comets, and asteroids crowded the inner solar system and rocked early Earth with numerous collisions, the infant Earth cooled and began to settle down. Unlike the decades-long perception that early Earth was a hot, volcanic mess, coated with numerous flows of lava, evidence now shows that the early history of the planet was dominated by a cool environment and with ample liquid ocean water.

Planetary scientists believe the scene turned somewhat ugly again during the so-called Late-Heavy Bombardment, a hypothesized period 4.1 to 3.8 billion years ago when scads of asteroids and comets rained in on our planet's surface, and on all the other young bodies of the inner solar system. This was long after Earth and the other planets accreted; but evidence for this period of bombardment exists from the Apollo Moon rock samples, which show a majority of melt-rocks on the Moon forming during this window. During this earliest, violent period of planetary formation, life on Earth was probably impossible.

But following the Late Heavy Bombardment, the story soon changed, and one day life arose on Earth. Our planet's age is thought to be about 4.54 billion years, based on the radiometric dating of meteoritic samples along with the oldest known Earth and Moon rocks. The oldest known Earth rocks, specifically, are those dated to 4.4 billion years from the Jack Hills region of Western Australia, betrayed by radioactive impurities in the zircon crystals they contain. The earliest microfossils known on Earth also come from Western Australia, from the so-called Strelley Pool formation, one of the oldest outcrops of sedimentary rock on the planet, and were discovered in 2011. They are primitive cyanobacteria, some 3.4 billion years old, and are the oldest known life we have. Spherical, oval, and tubular shaped, they span a mere hundredth of a millimeter across. Specimens that are more controversial could push the age back to 3.9 billion years, but they are as yet unconfirmed.

Scientists are just beginning to understand the complexity of how life arose on Earth, and how quickly it might have arisen. But more amazingly, they are also recently able to predict the outcome of life on Earth – to forecast how and when life on our fragile blue marble will come to an end. But before getting to the details of that fascinating – if a little depressing – outlook, it's important to understand what life is, how it may have started on our planet, and the basic timetable of how we got here.

What exactly constitutes life? Ideas have shifted substantially over the centuries. In ancient Greek Sicily, Empedocles (490 BC–430 BC) argued that everything is made of matter formed from earth, water, air, and fire, and that living things consisted of the same stuff as nonliving things, just in a different arrangement. Later, in Athens, Aristotle (384 BC–322 BC) proposed hylomorphism, the idea that all things are composed of matter and form and that living things have souls. Much later, in the seventeenth century, a variety of philosophers took up the idea of vitalism – that there is a fundamental difference between organic and inorganic matter. In 1828, however, German chemist Friedrich Wöhler (1800–1882) disproved the idea when he synthesized urea from inorganic materials.

In modern times, biologists agree on several factors that distinguish living things from ordinary matter. Darwinian evolution certainly plays a big role in life on Earth, and biologists see no reason why it also would not play a major role in life on other worlds in the universe. Life as we know it certainly begins

Figure 3.1a Comet Lovejoy appears to blaze through Earth's atmosphere in this striking image captured by International Space Station Commander Dan Burbank in 2011. Planetary scientists believe comets supplied the planet with building blocks essential to life several billion years ago.
NASA

Figure 3.1b Astronauts onboard the International Space Station captured this image looking across Earth's atmosphere, which blocks out harmful radiation from the Sun that would otherwise bombard our planet.
NASA

with organic (carbon-based) molecules and with water as a medium to carry energy and information (a solvent for chemical reactions), and a source of usable energy – the Sun.

It is useful to ponder a little basic chemistry when thinking about what constitutes life. Four essential types of molecules play key roles in living organisms: proteins, carbohydrates, lipids, and nucleic acids. Proteins are molecules that contribute the basic structural properties of life, and help to get functional jobs done, and are large polymer chains composed of amino acids. Carbohydrates are molecules consisting of carbon, hydrogen, and oxygen atoms, and they include sugars; carbohydrates are a source of energy. Lipids are molecules that provide cell structures, energy storage, and signaling, and include fats, waxes, sterols, some vitamins, and other molecules. Nucleic acids are complex molecules that form the buildings blocks of life's templates, including the genetic codes and self-replicating aspects of RNA (ribonucleic acid) and DNA (deoxyribonucleic acid), life's two most important complex molecules.

Astronomy enthusiasts often wonder about the roles of carbon and water in life on Earth, and whether all life in the universe would be related to them in the same way. Carbon is key to life on our planet for a variety of reasons – it is a reactive element, easily bondable to others, and it forms a variety of chain, sheet, and ring structures that play significant roles in biological molecules. It is also very abundant. But other elements could also play central roles in life on other worlds. Silicon, for example, bonds and also forms those life-friendly molecules in a similar way to carbon, but requires higher temperatures to break bonds. The similarities are responsible for creating fossils on Earth, when silicon replaces other elements in dead creatures. But silicon is not nearly as abundant as carbon in the universe. Still, silicon could serve the same purpose for other beings in the universe as carbon does to us.

Water is critically important as a solvent and a carrier of energy for several reasons. Oxygen is extremely abundant in the universe and one of its primary molecular forms is as water – the other is carbon monoxide. Water is extremely abundant on Earth because of the early period of bombardment our planet underwent by water-rich asteroids, comets, and by the accretion of water-rich grains of material that built our planet. Water as a liquid exists in a range of temperatures that matches well the daily business of organic chemistry – from 273 to 373 K. This is the case because water bonds in an unusual way.

The bottom line is that water serves as a medium through which biochemical reactions take place within all living things on Earth. Therefore, biologists believe that liquid water – or at least another similar liquid medium – is required for life's basic processes. Perhaps ammonia-water solutions, methane, ethane, or other hydrocarbon liquids could serve the same purpose at lower temperatures than those we have on Earth.

These biochemical speculations have led planetary scientists to propose that perhaps microbial life could exist not only on Mars – where we know subsurface ices and presumed subsurface liquid water exist – but perhaps on a range of more exotic worlds in our own solar system too. These include such exotic spots as the atmospheres of the giant planets, the liquid water oceans (underneath an icy shell) on Jupiter's moon Europa and on Saturn's moon Enceladus, and in the abundant methane rivers and lakes on Saturn's largest moon Titan.

This raises the question of whether, if we found microbial life on another world, we would be able to recognize it as life. Already, in 1996, planetary scientists put forth a martian meteorite discovered in Antarctica, ALH 84001, alleging that it contained fossils of microbial martian life. After extensive analysis, however, virtually the entire scientific community concludes the fossil-like structures are inorganic, although the debate and further research continue.

What exactly defines a living organism? Biologists generally agree on all or nearly all of seven traits necessary for life: homeostasis, the regulation of the organism's internal environment; organization by cells or other structures; metabolism by transforming energy from matter and utilizing it for work; growth over time; adaptation, the ability to change over time; response to stimuli; and reproduction, the ability to make completely new individual organisms. Adaptation is a very strong indicator of life – one that's hard to top.

So how did life get going on early Earth? No one yet knows the exact story, but biochemistry provides a tantalizing scenario. Certainly, scientists know that early Earth was richly seeded with the building blocks of life. Within hundreds of millions of years after our planet accreted, a watery, cool environment existed on Earth. Lots of complex compounds useful for living things were delivered by comets and asteroids in the form of ammonia, carbon dioxide, methane, carbon monoxide, nitrogen, and many others. The primordial soup was charged up by ultraviolet light and zaps of lightning, forming ever-more complex organic compounds.

It is clear from meteorites and comets that organic molecules up to the complexity of amino acids, the building blocks of proteins, are synthesized readily, as they exist in a variety of primitive meteorite samples and in the only comet to have been sampled thus far, 81P/Wild (Wild 2), in material returned to Earth by the Stardust mission in 2006. Interestingly, such complex molecules are chiral; that is, they can form in two flavors that are mirror images of each other. This plays a big role in the way they interact with other molecules.

With amino acids, for example, left-handed or right-handed varieties can interact in the same way with symmetrical molecules or with others of their own type, say, left-handed. But left-handed and right-handed amino acids do not interact in the same way. So the "handedness" of biological molecules is critical to their functionality. All proteins on Earth are constructed from left-handed amino acids, and RNA and DNA use only right-handed sugars. Life on other worlds, however, could be different, using right-handed amino acids, for example, and left-handed sugars.

That amino acids exist throughout planets, comets, and asteroids we can understand. But it is a big jump to go from amino acids to life. Biologists believe that life arose on early Earth through one of two mechanisms – "genetics first" or "metabolism first." Both models begin with small organic molecules synthesized on Earth or in meteorites or comets and delivered to Earth. In the "genetics first" model, an information-carrying molecule forms and eventually begins to control the chemical reactions in primitive cells. In the "metabolism first" model, networks of chemical reactions evolve toward higher complexity until they develop information-carrying molecules.

Whichever model was in play, eventually increasingly complex organic molecules evolved. Either vesicle factories – early cell analogs – evolved that eventually made molecules of higher and higher complexity, or RNA formed from abiotic chemical processes and was incorporated into the chemical factory vesicles. Biologists believe RNA preceded DNA because RNA plays central roles in all living cells. Evolving RNA from abiotic chemistry would not have been an easy trick, but it could have happened from existing nucleic acid bases, phosphate group chemicals that connect the bases, and the sugar ribose, which serves as a molecular binder to attach the parts together.

Biologists do not yet have a clear picture of the earliest life on Earth. But we know the basic chemistry existed that would allow for it, and it is tempting to think of it happening not only in the liquid water oceans we know existed on early Earth, but perhaps deep in the ocean, beside hydrothermal vents that are rich in organics and convenient sources of heat. Hydrothermal systems following impacts may also have been important incubators of life on Earth. Once large impacts on Earth slowed down, the formation of life on our planet may have happened relatively quickly. But no one knows how long it would have taken to form self-sustaining vesicles, the chemical factories, or RNA.

It could be said that the process was inevitable. Atoms are attracted to each other by electrical charges, set to combine in specific ways. That's why planets are built in certain ways and why minerals form with such high precision. The same holds for organic molecules, and preferred paths of chemistry for complex organic molecules, the stuff of life – and choices for particular metabolic cycles, may well exist in nature.

Once it was established, it appears that life took a mighty foothold on the planet at a rapid pace. The two most critical of life's components, carbon and water, exist in great quantity and can be extracted from the environment, as with carbon dioxide. It is quite possible that the RNA World commenced life. An RNA World suggests that self-replicating RNA molecules were the precursors of the life that followed. Or perhaps it was originally an Iron-Sulfur World of organisms metabolizing iron sulfide just like those earliest microfossils discovered in Western Australia. An Iron-Sulfur World suggests that life could have begun on the surfaces of iron sulfide minerals. Or perhaps life began as a Lipid World, in which double-walled bubbles of lipids like those in cellular membranes arose in the oceans. A Lipid World suggests that life began as a self-replicating object that resembled a lipid; a chemical class of molecules that contains fats, waxes, sterols, and other complex molecules.

However it arose, life began as simplistic proto-bacteria and stayed that way for a very long time. (Most scientists believe that viruses, discovered in 1892 and thought to be alive in the early twentieth century, are really non-living predators that attach themselves to living cells and overtake them, so are not independently living. But some argue that viruses are alive because they can adapt by natural selection.) The first 1.5 billion years of life on Earth featured nothing more than simple prokaryotes, primitive microbes lacking a cellular nucleus. Some 2 billion years ago the first eukaryotes appeared – microbes with a cellular nucleus. Multicellular organisms appeared over a vast range of time. Over time, a buildup of oxygen in Earth's atmosphere made many other forms of life possible in the coming eons. Less than a billion years ago, invertebrates came onto the scene.

Within the last half billion years, Earth's climate and its set of living organisms underwent dramatic change. The Cambrian period, beginning 541 million years ago, saw an explosion of the first large numbers of mineralized (and thus fossilized) large multicellular organisms such as trilobites and brachiopods. During the Ordovician period, beginning 485 million years ago, marine invertebrates, including jawless fish, prospered. The Silurian period, beginning 444 million years ago, brought the first jawed fish and the first terrestrial arthropods. During the Devonian period, beginning 419 million years ago, fish

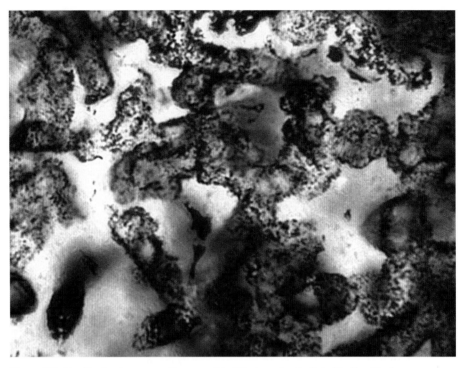

Figure 3.2 Fossil microbes were discovered in Western Australia's Strelley Pool, an area thought to have been one of Earth's first beaches. The tubes seen here are about 10 microns across, about the size of a small piece of pollen.
David Wacey

diversified and early amphibians appeared. The Carboniferous period, beginning 359 million years ago, saw the appearance of reptiles. And the Permian period, beginning 299 million years ago, saw commonplace mammal-like reptiles and widespread extinction of amphibians.

Then a great transformation on Earth occurred during the Triassic period, which began 252 million years ago. The mammal-like reptiles met their end and were replaced by dinosaurs, creatures that by rights should have existed down to this day, if not for a certain earthbound asteroid. The Jurassic period, beginning 201 million years ago, brought the first birds and great diversity among reptiles. Starting 145 million years ago, the Cretaceous period was warm and wet – a greenhouse period – until its end, when ancient birds and reptiles met their extinction.

Then came the Cretaceous-Paleogene Impact (K-Pg Impact) that killed off the dinosaurs and commenced the Paleogene period, 66 million years ago. The modern orders of mammals evolved further. Human beings evolved over the past 5 or so million years. Our closest ancestors evolved over the past 2 million years and led, starting roughly 200,000 years ago, to *Homo sapiens*. And here we are.

When I talk to astronomy groups about life on Earth, and how incredible it is that we can be part of the universe and be *thinking about* the universe around us, some surprising things often come up. "How long will life on Earth last?," I'll often ask. The first thing knowledgeable, interested astronomy enthusiasts think of is that the Sun is about halfway through its life, as we have just seen in the previous chapter. So life on Earth must be about halfway through our journey, too, right? That notion, which we have all seemingly had for years, couldn't be farther from the truth.

Over the last 20 years, a number of studies have looked at this intriguing question. As we have seen, the Sun is a variable star, and its total bolometric magnitude – the total amount of radiation it gives off – is slowly but surely increasing. One of the earliest and most influential was a 1994 study by New York University biologist and geologist Michael Rampino (1948–) and atmospheric scientist Ken Caldeira, then of the Lawrence Livermore National Laboratory, who summarized a broad survey of research in the *Annual Review of Astronomy and Astrophysics*.

The Rampino–Caldeira study cast a major focus on many overlapping areas of great interest, all pertaining to the future of life on Earth. Titled "The Goldilocks Problem: Climatic Evolution and Long-Term Habitability of Terrestrial Planets," it served as a tour de force in forecasting Earth's future for living beings.

This study commenced by asking why Venus is too hot, Mars too cold, and Earth just right for life. Distance from the Sun might seem to be the answer, but there is much more to the story than that. To really understand the Goldilocks Problem, you need to understand the early history of planetary atmospheres, and also be able to forecast their evolution. The long-term evolution of the Sun is, of course, the central player in this drama.

Planetary scientists know that early Earth was warm enough for liquid water to exist, despite the fact that the Sun was some 25 or 30 percent fainter back then. How do we know liquid water existed? It deposited sediments. Early Earth probably had a lot more carbon dioxide in the atmosphere than it does now, or maybe methane, which was taken out by the growth of oxygen from photosynthesis.

And the stakes were high. Had Earth been a little colder, it might have permanently frozen over and life, at least as we know it, may never have evolved. But scientists know that liquid water has existed on Earth for at least 3.8 billion years, and probably more. The production of carbon dioxide might have averted a potential ice-over on Earth. But could too much carbon dioxide also halt the weathering of rocks that removes carbon dioxide from our atmosphere and produce a runaway greenhouse effect that would transform our atmosphere into something like that in Venus?

By all rights, the effective surface temperature on Earth should be below the freezing point for water. But it is actually warmer through the greenhouse warming by water and carbon dioxide in our atmosphere. The mean surface

Figure 3.3 Nearly 4 billion years ago, the young Sun's energy was far less than it is today. Nonetheless, we know liquid water and life existed even then.
Ron Miller for *Astronomy*

temperature of Venus is 750 K (too hot), Mars 218 K (too cold), and Earth 288 K (just right), not only because of greenhouse gas moderation, but also from distances from the Sun and the percentage of reflectivity of the planets – and Venus is highly reflective or it would be much hotter.

Earth and Venus also show huge differences in their "handling" of carbon dioxide, despite the fact that they received about the same amount of the gas, by whatever means – bombardment by impactors, outgassing from the interior, and so on. On Earth, at the surface, much of the carbon dioxide was removed by the carbon cycle and locked up in carbonate rocks. Venus was always too warm for liquid water, and hence had no weathering process to remove carbon dioxide.

So how did Earth avoid being too hot or too cold for life over the entire history of its existence, for the better part of the last 4 billion years? Consider that Earth is unique among terrestrial planets for containing an enormous quantity of water, much more than might exist at this distance from the Sun. Our planet's surface, or near it, has 1.4 billion cubic kilometers of water, 97 percent of which lies in the oceans, with some in the polar ice caps and glaciers. Water also exists under the surface, although how much is not exactly known. Internal water may exist because of the way Earth's surface melted during the planet's accretion, allowing water to be absorbed by molten rock. The planet's surface water alone amounts to a layer 2.7 kilometers thick over the entire globe.

But Earth is not in a completely stable state, both because of what's happening with the Sun's slowly increasing radiation and due to many changing factors on our planet itself. The fate of life on our planet is unclear, but we can look at one absolute, the increase in solar radiation. There will come a time when Earth is completely uninhabitable, all its life killed off, because the temperatures are too high and the oceans have boiled off. (Although it must be said that microbes inside Earth's crust will be able to outlive everything on the surface and in the oceans.) And unlike the notion that we should be about halfway through life on Earth, analogously to the Sun being roughly halfway through its life, we may actually be closer to the end than anyone would have guessed a generation ago.

A key question in foretelling the end of life on Earth relates to the evolution of greenhouse gases on our world. In the early 1980s, several groups of

Figure 3.4a Dust storms push across the Red Sea from Sudan into Saudi Arabia. If humanity survives into the distant future, twenty-first-century climate change will seem like a minor inconvenience compared to the challenge of a swelling Sun.
NASA

Figure 3.4b Lake Mead's water supply from the Colorado River is dwindling, threatening the long-term prospects for neighboring Las Vegas.
NASA

scientists proposed that atmospheric carbon dioxide could be nearly depleted some 100 million years from now, due to chemical weathering of rocks. Over time, the increase in solar radiation is compensated for by the drawdown in carbon dioxide. But we can't go much lower in atmospheric carbon dioxide or plants won't be able to photosynthesize. The feedback loop will no longer work.

Photosynthesis, a critical process in organisms, extracts carbon from the atmosphere and from dissolved carbon dioxide, but will ultimately be disrupted by increased solar luminance, resulting in a lack of carbon available in the biosphere. Unless organisms change so they can exchange carbon dioxide with the atmosphere more efficiently, this may happen in less than a billion years. If photosynthesis stops, then there will be no production of calcium carbonate in the oceans and hence less carbon dioxide coming from volcanoes.

One factor will be working in the opposite direction. Subducting carbonate rocks on the ocean floor, sliding into the heated mantle, will significantly increase the amount of carbon dioxide they release into the atmosphere in future times. But this will not have a huge effect, and even as Earth warms significantly, the oceans will likely maintain a balance between silicate and carbonate rocks, also roughly maintaining their pH, and staying around for a long time to come. However, a warmer Earth will eventually mean much more carbon dioxide, and this is independent of any global warming created by humans as part of our industrial lifestyle.

Where does all this lead? Planetary scientists know that regardless of other factors, the Sun's increased energy output will fundamentally change Earth over the coming few hundred million years. Earth's global surface temperature will likely rise to 80 °C (176 °F), approaching the boiling point of water, in 1.5 billion years. But with more carbon dioxide in the atmosphere that point could come sooner. Based on the relationship of Earth's surface temperature to water mixing ratios in the stratosphere, calculations show that Earth's oceans could be gone on a timescale of a billion years or less. And the last molecules of water on the planet could be gone within 2.5 billion years. That would be the end game for life on Earth. But for the more fragile types of creatures, including human beings, life on Earth would cease well before the billion-year loss of the oceans.

As we have discovered, the most ancient fossil life found on Earth thus far is 3.4 billion years old, those cyanobacteria microfossils from the Strelley Pool in Western Australia. (Some scientists argue there is chemical evidence for life dating back 3.9 billion years.) If we take a reasonable guess at the end of life on Earth, or at least the vast majority of life, at 800 million years from now, then perhaps 80 percent of the story of life on Earth is already written. When Earth's water is gone, silicate-rock weathering will stop and there will be a runaway surge of carbon dioxide in the atmosphere, leading to a Venus-like result. Our blue marble will then transform into a hellish, unrecognizable, and very hostile world.

What can humans do, if we are still around, to survive that inevitable holocaust? For almost a generation, some planetary scientists have proposed

Figure 3.5 A swollen red giant Sun fills Earth's scorched horizon billions of years from now, spelling doom for the life that might remain on it.
Ron Miller for *Astronomy*

terraforming worlds, that is, transforming a planet like Mars to be more habitable than it currently is. Among many, Chris McKay of NASA's Ames Research Center has made extensive studies of what could be done with the martian atmosphere. The Red Planet is by far the most looming candidate for terraforming, as it is the most earthlike, is close by, and would offer a possibility of atmospheric alteration. Early studies of the martian atmosphere suggested that reengineering the levels of carbon dioxide in Mars's atmosphere could raise the planet's surface temperature above the freezing mark. Covering the polar caps with dark material, for example, could release significant quantities of carbon dioxide into the planet's atmosphere. But those studies misinterpreted the nature of the polar caps, which actually volatize and are redeposited every year.

More recent studies by McKay and others suggest that if large amounts of volatiles exist underneath the martian surface, then schemes could be employed to release large amounts of carbon dioxide and water from surface rocks. But exactly how much carbon dioxide is locked up in carbonate rocks on the surface is not well known. And, of course, the scale of such an engineering project on another world is staggering – in terms of spacecraft, monies, effort, and everything else – at best to contemplate.

The question can also be asked, "Will humanity even make it another 800 million years to see the final possibility of life on Earth?" Lots of other dangers exist in space and on our planet that could stop the grand human experiment well before then. To underscore the violence of the universe, we can look back at a few examples of extinction events in our planet's history, of which we know plenty about.

Of course, the most famous mass extinction event in Earth's history is the K-Pg Impact, formally known as the Cretaceous–Paleogene Extinction Event (and formerly known as the K-T Impact, for Cretaceous–Tertiary), the most celebrated of the end-Cretaceous events. It was caused by an asteroidal impact in the Yucatán Peninsula in Mexico some 66 million years ago, and which, among other things, killed off the dinosaurs. In 1980, physicist Luis Alvarez (1911–1988) and his son geologist Walter Alvarez (1940–) hypothesized the impact–extinction scenario, and scientists identified the smoking gun, Chicxulub crater, in the Yucatán Peninsula, in 1990. The K-Pg event killed 17 percent of all living families, 50 percent of all genera, and 75 percent of all species on Earth. This event enabled mammals and birds to emerge as dominant in an age of renewed life.

The causes of mass extinctions are many. They include changes in sea level, global warming, global cooling, marine anoxia (lack of oxygen for sea creatures), changing ocean–atmosphere circulation patterns, increasing solar radiation, plate tectonics, Large Igneous Province volcanism (eruptions from special geological hotspots), and large comet or asteroid impacts.

Many extinction events have taken place in Earth's history, and five, including the K-Pg extinction, represent the most acclaimed. The others in the hall of fame are the Triassic–Jurassic extinction event, or the end Triassic, some 201 million years ago. This killed 23 percent of all families, 48 percent of all genera, and 75 percent of all species. This eliminated many of the chief competitors of the dinosaurs.

The Permian–Triassic extinction event, end Permian, happened some 252 million years ago. This is the largest mass extinction event known in our planet's history, and killed 57 percent of all families, 83 percent of all genera, and as much as 96 percent of all species. This so-called "great dying" event fundamentally changed the course of life on Earth.

The Late Devonian extinction event, roughly 375 million years ago, consisted of a series of extinctions that reduced the number of families by 19 percent, the number of genera by 50 percent, and the number of species by 70 percent. This series of events may have lasted for some 20 million years.

The Ordovician–Silurian extinction event, end Ordovician, occurred about 444 million years ago and consisted of two events that killed 27 percent of all families, 57 percent of all genera, and as much as 70 percent of all species.

So life on our planet is a pretty fragile thing, and just because we know about the extinction events does not mean they will simply stop in the future. There is an old, silly saying – "Mother Nature is a bitch!" When the universe seems stacked against you, when everything seems to be going wrong, sometimes that old joke resurfaces. But in reality, nature is merely ambivalent – or, more properly, nature cares for nothing and no one except for its own laws.

The current global warming on Earth, accelerated by man-made causes (a scientific fact, by the way, supported by virtually 100 percent of climate scientists), has started us on a new and profound experiment. No one knows as yet where this will take humans in the future. (Who knows? It may help or have helped us avoid a potentially much worse ice age!)

And there are other dangers lurking out there, far away from our own planet. We know about impact hazards. Many large impacts have occurred in Earth's history. Some 12,000 Near-Earth asteroids or comets have been discovered and catalogued, and many others certainly exist. Some are on orbits that eventually will intersect Earth's orbit, unless something is done to deflect them. Again, our awareness of this situation is good, but does nothing to stop the problem. On average, a 1-kilometer asteroid should strike our planet about once every 700,000 years. A 140-meter asteroid will strike Earth on average every 20,000 years. The impact interval of an asteroid a few meters across, like the rock that fell into Chelyabinsk, Russia, in 2013, is roughly every 50 years. Will we be ready to ward off the next one?

More distant trouble could come from nearby parts of our Milky Way Galaxy, too. A nearby nova or supernova could pose a huge problem because of sterilizing radiation, and a gamma-ray burst could do this from a much greater distance. And much less likely, but plausible events could pose dangers from afar too: what about mini black holes, or antimatter encountering our solar system? In those cases, we would never know what hit us. These possibilities are extremely unlikely, but they are nonetheless possibilities. On shorter time-scales, coronal mass ejections from the Sun might pose a more significant risk.

If humans do survive in the long run, our species – and all the other life on Earth – will have interesting signposts ahead. There is also the huge issue of limited resources on Earth, not just in terms of food but also in the key necessities of technology and life-sustaining elements. Will overpopulation create a vast crisis over resources in a multitude of ways?

Whether or not humanity survives, the universe has interesting changes ahead. In just 100,000 years, the proper motions of stars will mean that most constellations we now know will no longer be recognizable.

In a million years from now, perhaps the red giant star Betelgeuse, the bright shoulder of Orion, will become a supernova, brilliantly shining in our daytime sky from a distance of 650 light-years. In 8 million years from now, Mars's potato-shaped moon Phobos will pass close enough to the Red Planet to be broken up into a ring of debris. Some 10 million years hence, on Earth, the Great Rift Valley will widen to the point that it will split Africa, creating a new ocean basin. Some 600 million years from now, the Moon will nudge far enough away from Earth such that total eclipses will no longer take place. And back to that number of 800 million years, when the Sun's increased luminosity will have stopped photosynthesis, all plant life will have been killed, disrupting the silicate–carbonate cycle. Earth will now change dramatically. Water will evaporate on our world's surface, rocks will harden, and plate tectonics will grind to a halt. As we lose our water, there will be no way to remove carbon dioxide emission from volcanoes, no weathering mechanism to balance the atmosphere. Therefore, we will have a very hot planet indeed, with solar luminosity rising.

As the temperature rises, carbon dioxide levels will fall to very low levels. Perhaps pools of water will remain for a time here and there, at the poles, and most of the planet will be a "moist greenhouse," drying away year by year. Simple life may still exist where small amounts of water remain. Earth's moist greenhouse will rapidly become a dry greenhouse, as water vapor rises to the upper atmosphere and escapes into space.

We will be gone – died off, or moved to another world, all of our memories of the dreams, glories, and accomplishments of humanity cast into the unforgiving, unrelenting history of the cosmos.

Chapter 4
How the Moon formed

When we walk outside at night and look skyward, it's usually the Moon, our planet's natural satellite that first catches our eye. The fifth largest moon in the solar system – after Jupiter's Ganymede, Callisto, and Io, and Saturn's Titan – it is the second moon in the solar system in relation to its parent body, Earth. (Charon is the largest moon in the solar system relative to its parent body, Pluto.)

The Moon's name originated from the Old English *mōna*, which arose before 725 AD, and followed a still older proto-Germanic name. Prior to that, the names for the Moon derived from the Latin *luna* and from the Greek *selene*. (A quick aside on nomenclature: note that Earth's name is *Earth*, not *the Earth*, just as you wouldn't say *the Saturn* or *the Jupiter*. But the Moon's proper name is *the Moon*, differentiating it from many other moons in the solar system. If you catch someone saying, "I'm glad to live on the Earth," tell them to stop it. There is enough foolishness going on in this world already. In addition, the media also often use a lowercase *m* for the moon. The proper name is the Moon.)

The Moon is a pretty innocuous body, going about its business, orbiting Earth, and showing its phases each month, from the dark "new," through the "first quarter" to "full" to "last quarter," wreaking havoc only to deep-sky observers who wish the Moon would go away and not spoil their dark-sky views of galaxies, clusters, and nebulae. The Moon stretches an impressive 3,476 kilometers across at the equator, which means you could just about wedge its width between New York and Phoenix.

Certainly, the Moon's "face" has been a familiar sight in the sky for the whole history of humankind. It is tidally locked with Earth such that, although it wobbles slightly, only one face covered with darkish maria ("seas") is visible to us. Even our earliest ancestors must have looked skyward night after night (or day after day!), wondering what this bright and changing light in the sky was. Amazingly, however, we have come to terms with how the Moon must have formed only in the last generation.

For decades, the Moon has presented planetary scientists with a variety of puzzles. To explain its origin, we need to account for the high angular

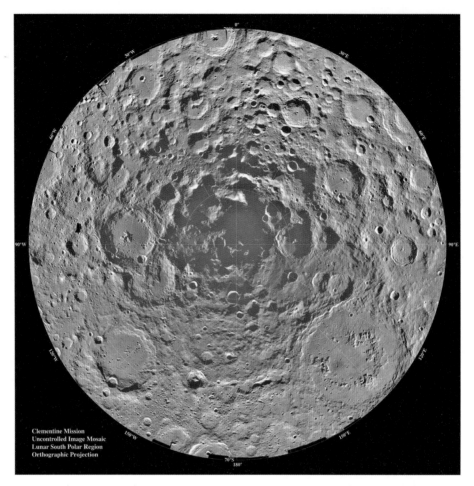

Figure 4.1 The Moon shows an unfamiliar face in this south polar view captured by the Clementine spacecraft.
NASA

momentum of the Earth–Moon system – that is, the high degree of rotational speed, orbital speed, and mass in the system. There is also the oddity of the Moon's orbit, which is inclined 5.1° from the plane of the ecliptic, the "disk" in which the planets orbit the Sun. In addition, the Moon contains a large amount of mass compared to Earth itself – the Moon has slightly more than 1 percent of Earth's mass, which is high relative to other solar system moons. The Moon also has a relatively low density compared to terrestrial planets, about 3.3 g/cm³, which is much less than Earth's density or those of the other inner planets.

And the lunar oddities don't end there. Beginning in 1969, and extending through 1972, Apollo astronauts returned 2,415 Moon rocks, some 382 kg (842 lbs.) of samples. Three Soviet Luna missions returned an additional 0.32 kg (0.7 lb.) of Moon rocks between 1970 and 1976. And more than 120 lunar meteorites are now known to have fallen on Earth and been collected. All these have given planetary scientists a rich and diverse suite of lunar rocks to analyze.

The returned samples added to the mystery of the Moon's origin. They are chemically complex, and their ages and the isotopes (elemental variants) they contain perplexed planetary scientists because the Moon rocks are relatively exotic compared with other worlds in our solar system – Earth included.

So where did the Moon come from? For most of human history, no one had even a clue. And it's still a puzzle we're working on with some very good guesses, but even now not with certainty. Historically, lunar studies were primitive for a very long time. But by the fifth century BC, Babylonian astronomers deciphered the Moon's Saros cycle, and understood the 18-year relationship between eclipses. Around the same time, the Greek philosopher Anaxagoras (ca. 510 BC–ca. 428 BC) worked out that both the Sun and Moon were huge, spherical rocky bodies, and that the Moon reflected sunlight. The great Greek philosopher Aristotle (384 BC–322 BC) reasoned that the Moon is a boundary between the spheres of the elements – earth, water, air, and fire – and the ether that lay beyond.

Understanding of the Moon came slowly in antiquity. In the second century BC, Hellenistic philosopher Seleucus (*flourished* 150 BC) was the first to argue that ocean tides result from the Moon pulling on Earth. Greek astronomer Aristarchus of Samos (ca. 310 BC–ca. 230 BC) and later Greco-Roman astronomer Claudius Ptolemy (ca. 90–ca. 168) calculated values for the size and distance of the Moon. The estimates of both Aristarchus and of Ptolemy were reasonably close to correct.

Skywatchers during the Middle Ages generally believed the Moon was a smooth, rocky body. But that illusion was shattered in late 1609 when Italian mathematician and astronomer Galileo Galilei (1564–1642) climbed to the rooftop of his house in Padua, Italy, with his recently constructed telescope, and swung his instrument focused on the nearby church steeple over to the Moon's disk. This first widely publicized telescopic astronomical observation not only was a crucial point for the entirety of modern science, but it instantly reworked what humans thought about the Moon. Our satellite was not smooth, but cratered, pocked, splattered – imperfect. (Historical purists know that Englishman Thomas Harriot [ca. 1560–1621] made drawings of the Moon as seen through a telescope some 4 months before Galileo did. However, he did not publish his work. Arguably, his full Moon map is better than any of Galileo's drawings.)

For several centuries, astronomers armed with telescopes carefully mapped the Moon, precisely drawing every feature they could find. Two early Moon mappers, Italian astronomer-priests Giovanni Battista Riccioli (1598–1671) and Francesco Maria Grimaldi (1618–1663), created the system we still use to name lunar features. In the nineteenth century, Prussian astronomers Wilhelm Beer (1797–1850) and Johann Heinrich von Mädler (1794–1874) created beautifully accurate renditions of lunar features.

Figure 4.2 In 2010, the Lunar Reconnaissance Orbiter spacecraft crashed a projectile into the Moon and then captured this image of the crater during its analysis of the ejected material. NASA/GSFC/Arizona State University

But the question of lunar origin still lay on the table. Well into the twentieth century, when planetary scientists were coming to realize that most lunar craters were probably impact related (rather than volcanic), the origin problem persisted. The possibilities were several: the Moon could have been captured from another orbit; Earth and the Moon could have formed simultaneously as a "double planet"; Earth could have given birth to the Moon in an episode of fission as it rotated rapidly; planetesimals, small bodies generally accreting into larger ones, coming into the area of Earth could have fragmented, creating the Moon; or a large impact could have taken place, somehow creating an Earth satellite. Three of these ideas were important in the pre-Apollo days, before about 1970: fission, capture, and co-accretion. The other two ideas followed the first Apollo results.

Thinking about the origin of the Moon began to pick up steam in the decade of the 1960s. In the Soviet Union, planetary scientist Victor S. Safronov (1917–1999) was a pioneering thinker about how planets form, having concocted the low-mass model of planetary formation. Safronov had produced studies about the accretion of planets from numerous planetesimals. The Apollo and Luna programs were underway, and lots of people were thinking about the Moon and associated questions.

In 1975, two planetary scientists at the University of Arizona's Lunar and Planetary Laboratory published a landmark paper proposing a novel idea for the Moon's origin. William K. Hartmann (1939–) and Donald R. Davis produced an intriguing hypothesis for the Moon's origin, which they published in the journal *Icarus*. According to Hartmann and Davis, a Mars-sized planetesimal, subsequently named Theia – Greek mythological mother of the Moon goddess Selene – collided with Earth, knocking out a considerable amount of debris, much of which formed a ring around Earth and eventually accreted into the Moon.

Hartmann and Davis studied the work of Safronov, focusing on a period about 4.5 billion years ago, and modeling the growth rates of bodies around the newborn and still accreting Earth. They compared the region to the main asteroid belt of today, wherein a dwarf planet exists, Ceres, with a diameter of just under 1,000 kilometers, and several bodies in the range of 500 kilometers in diameter. Hartmann and Davis reasoned that in a similar fashion, the region of the still-accreting Earth would have had several bodies of similar size, about half that of Earth, and that an impact with the hypothesized planetesimal happened late enough in the process of Earth's accretion that it cast out substantial material from the interior of Earth to form the Moon.

As with many unusual new ideas, the notion at first failed to catch on. Hartmann and Davis had actually mentioned the idea at a conference on planetary satellites in 1974, at which time another researcher, Harvard astrophysicist Alastair G. W. Cameron (1925–2005), revealed that he and a fellow scientist were also working on a similar idea, driven by the unusual aspects of the Earth–Moon system's angular momentum. A small degree of momentum for the idea caught on when not only Hartmann and Davis published their results, but also Cameron and his colleague William Ward published their idea in 1976.

By now, the idea had taken on the name Giant Impact Hypothesis but it had little traction among most of the planetary science community. (To be fair, most of the planetary science community does not focus on lunar origins.) The early 1980s saw some intermittent computer modeling and interest in the idea, but it was the major conference in Kona, Hawaii, in October 1984, 'Origin of the Moon', and the resulting publication in 1986 of the conference proceedings, edited by Hartmann and colleagues G. Jeffrey Taylor and Roger Phillips, that turned the spotlight onto this idea of lunar formation.

This meeting sparked a cottage industry among lunar researchers to investigate the Giant Impact Hypothesis, and many scientists increased their time spent on the problem. Numerous computer models were created to look at the interaction between the two large bodies and the resulting state of the debris. How much material would have been blasted into the space surrounding Earth? How much of the impacting body would have been absorbed by Earth? How long would this process have taken?

The idea received a rebirth in the mid-1990s, when Robin M. Canup (1968–), now a scientist and administrator at the Southwest Research Institute in Boulder, Colorado, received her Ph.D. Canup produced a re-analysis of all of the salient factors. In 1997, Canup revealed her study and how it suggested that such a huge impact most likely would result in a series of moonlets rather than one large Moon. However, subsequent studies, including a substantially detailed and important paper published in 2003, refined her work and led to results that seem relatively consistent with what we see in our sky.

But just how did we get to the point of taking the Giant Impact Hypothesis so seriously? The formative ideas that came about early on in the twentieth century were not all clearly defined, and aspects of some appeared to crop up in others. In order to weigh the evidence, it's worth taking a look at the basic properties of the most popular previous ideas.

For example, the model of intact capture of an already existing Moon from another orbit was advocated by some. But this idea has been shown to be highly unrealistic on the grounds of orbital dynamics. It does nothing to address the strange composition of the Moon, especially its clear relation to Earth as shown by the oxygen isotopes. If the Moon had simply been captured, planetary scientists would expect it to resemble a more run-of-the-mill primitive early solar system moon, like some of the satellites of the outer planets, which have a rocky and icy composition.

In fact, when American chemist Harold C. Urey (1893–1981) examined the Moon problem, he latched onto the fact that the Moon's density was similar to that of carbonaceous chondrite meteorites, and predicted that the Moon was probably a captured satellite because of this fact. But this was before we knew enough about the Moon's composition to rule that out. (And on a personal aside, it was Urey who headed the Manhattan Project group at Columbia University during World War II that included a young chemist named John H. Eicher – my father.) When it became clear that the Moon's composition was unusual and unique, then the capture idea plummeted in credibility, as it would be extremely unlikely to think such an unusual object was captured from a preexisting orbit.

Then there was the idea that Earth and the Moon formed in situ as a "double planet" system, so called because of the Moon's large size relative to Earth. But this idea runs into severe trouble on the grounds that the densities of the two bodies are so wildly different. In terms of composition, the Moon is not all that unusual. It is simply depleted in bulk iron compared to Earth. In fact, the bulk Moon is nearly identical in density and chemistry to uncompressed Earth upper mantle. Despite the inconsistencies, the double planet idea was so attractive, so basically logical, that planetary scientists investigated it for many years with a wide range of modeling.

To get around the problem that the two bodies have very different densities, scientists modeled accretion of the two bodies involving a ring of silicate rich debris with a low density. Imagine that planetesimals broke up in our vicinity, and in this furor created a silicate-rich ring around the proto-Earth, a ring with a pretty low density. Earth and the Moon could have formed in a differentiated way, incorporating various amounts of the low-density silicates. The higher density metallic cores of the planetesimals would have ended up mostly in Earth's core, and the mantles of the broken up planetesimals, lighter and silicate rich, could have formed most of the Moon.

But there is a significant problem with this idea. The breakup of planetesimals close to the proto-Earth, which would have to occur to produce this scenario, is extremely unlikely to have taken place. On dynamical grounds, it is highly unlikely to achieve the necessary angular momentum in such a scenario. In the unlikely event that such dynamics could have been happening in the early solar system, then this process of breaking up protoplanets and reaccreting them should have been common, and scientists might expect to also see Venus with a substantial moon (it has none).

The next possibility came as early as 1879, when the son of the famous Charles Darwin, English astronomer George Darwin (1845–1912), proposed that the Moon might have arisen from rotational fission – that is, that the Moon formed from material cast out of a rapidly rotating Earth. Early on, this idea and variants were popular because they could explain a Moon existing with a low density like the one we observe, and also a metal-poor composition. But this model does not explain the high angular momentum of the Earth–Moon system; the rotation rate for early Earth was not nearly high enough to have cast out material to make a Moon.

Apollo samples returned to Earth gave planetary scientists the first opportunity to test these models in a significant way beginning in the 1970s. In the Moon rocks, the isotopes of various elements are key – that is, the elemental variants, dictated by the numbers of neutrons they contain. Lunar samples show the same isotopes of oxygen and chromium as in Earth rocks, but enstatite chondrite meteorites, originating from asteroids, also have the same oxygen isotopes.

Earth and the Moon have significant isotopic and compositional differences. The Moon contains far more iron oxide than does Earth, and it also has lots more elements like aluminum and uranium and fewer elements like bismuth and lead. Similar abundances in vanadium, chromium, and manganese in Earth and the Moon are also shared by many chondrite meteorites. Scientists also proposed that perhaps the fission hypothesis would work if multiple small impacts had ejected mantle material from Earth into orbit, to be accreted into the Moon. But the extraordinarily high angular momentum for this is almost impossible to believe. And chemical differences between Earth's mantle and the Moon did not help the idea that the Moon arose from Earth.

Moreover, if these ideas were credible, other Moon-like satellites might have been produced around the other terrestrial planets, which, of course, they haven't. None of these ideas addresses the unique system we have with Earth and the Moon, the similarities and differences between the bodies, specifically the differences between Earth's mantle and the Moon. They do not explain the Moon's orbit or the high angular momentum of the Earth–Moon system. Something much more profound and amazing was clearly needed.

When Hartmann, Davis, Cameron, and Ward published their lunar origin idea in the 1970s, they focused largely on solving the problem of this high angular momentum in our system. But their work began to solve other problems too. Let's consider what it really says. During the final stages of the accretion of the planets, a Mars-sized body slammed into Earth, spinning out a disk of debris that formed the Moon. Cleverly, the Giant Impact Hypothesis solves quite a number of problems associated with how the Moon formed.

The process may have worked something like this. We know that Earth is about 4.6 billion years old. Some 50 or 100 million years after Earth's accretion, our planet suffered a grazing impact with a Mars-sized body having about 10 percent of Earth's mass. (The body took on its name, Theia, only in 2000, at the suggestion of English geochemist Alex Halliday.) Scientists believe the body consisted of a metallic core and a rocky mantle, and originated from the same area of the protosolar nebula as Earth, and therefore would have had a low enough impact velocity to produce the Earth–Moon system, and also would have had the same isotopic ratios of oxygen and chromium as Earth.

Heavily damaged and disrupted by the collision, most of Theia's rocky mantle would have settled into orbit about Earth. Because it had been slammed so hard, Earth was shaped asymmetrically for a time – like an egg – and that created torques that helped to suspend material into orbit. Gases rapidly expanded from the destroyed parts of Theia, also accelerating material into orbit. The material that had existed in Theia's mantle was shocked and mixed and flung at high speeds into a disk, whereas the impactor's metallic core remained more stable, falling into Earth within a few hours, and being absorbed into our young planet. The low-density ring of silicate material from Theia's mantle was suspended in orbit.

The next steps in the process vary from model to model. Some scientists imagine a molten Moon accreting quickly and others envision a series of moonlets accreting, which came together to form a molten Moon over a longer interval of time. Cameron's initial model suggested an impact event with differentiation of the impactor's core and mantle within a minute of the event, a blending of much of the material after 10 minutes, a string of mantle suspended in orbit after an hour, the impactor's core being reabsorbed into Earth after 4 hours, and an orbiting ball of proto-Moon debris being left after 24 hours. However it happened, the event between Theia and Earth took place remarkably quickly – most of the action lasting a day, and the resulting absorption of most of Theia into Earth and the formation and subsequent evolution of the Moon lasting for billions of years to come.

The giant impact clearly vaporized a large fraction of the material that existed in Theia, but the Moon did not accrete from vaporized solids. The fact that the Moon is extremely dry may be a reflection that Theia was extremely dry, and in fact most of the inner solar system was and is extremely dry. Earth, with our abundant water, is the clear exception. Most of Earth's water is believed to have come from bombardment by water-rich asteroids, comets,

Figure 4.3 Planetary scientists believe the Moon formed in a shattering impact between Earth and a planetesimal called Theia, some 4.53 billion years ago.
NASA

and even by the adhesion of water molecules to grains that accreted to form our planet, and perhaps many of the water-rich objects that struck us came from the region of Jupiter or beyond.

The Theia hypothesis requires that a really large impactor existed in the earliest days of the solar system and was on a trajectory to intercept and strike into Earth. There is no doubt that the inner solar system was filled with far more impacting bodies in those days than it has been since. But do we really know that a body as large as Mars could have existed or did exist in Earth's vicinity that long ago? (It is thought to have been a low-velocity impact, so Theia could have been at the L_4 or L_5 Lagrangian point for a while.)

Evidence of large impacts exists elsewhere in the solar system too – in the inner and outer solar system. Large impact basins such as the South Pole–Aitken Basin on the Moon's far side, which spans 2,500 kilometers across, are indicative of huge impacting bodies. The Caloris Basin on Mercury stretches 1,550 kilometers across, while on Mars the Utopia basin spans an incredible 3,300 kilometers, the largest impact basin in the solar system. Moreover, huge impacts are known to have occurred farther out too, as with the impact between

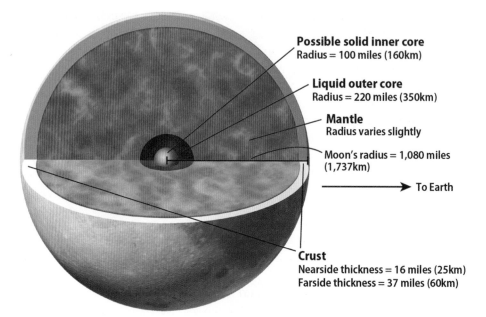

Possible solid inner core
Radius = 100 miles (160km)

Liquid outer core
Radius = 220 miles (350km)

Mantle
Radius varies slightly

Moon's radius = 1,080 miles
(1,737km)

→ To Earth

Crust
Nearside thickness = 16 miles (25km)
Farside thickness = 37 miles (60km)

Figure 4.4 After the impact between Earth and Theia, planetary scientists believe the Moon settled into a body with an interior superficially similar to Earth's. However, the Moon's core is tiny in comparison and the crust is much thicker.
Astronomy: Roen Kelly

an Earth-sized body and Uranus, which re-oriented the spin axis of the giant planet on its side.

Because of these examples and others, it seems clear to planetary scientists that many very large impacts did take place in the early solar system. It seems not only believable but also likely – considering what is known of the early solar system – that a large single impact from a long-gone body such as Theia might have created the Moon.

But that's not the whole story. Research has continued beyond the formative years of the hypothesis. In 2001, planetary scientists at the Carnegie Institution of Washington produced a paper with very high precision measurements of the isotopes within Moon rock samples. They found Apollo samples to be identical in significant ways with Earth rocks, and very different from those of other solar system bodies. The idea that most of the material in the Moon came from Theia took a step backward, and in 2007 planetary scientists from the California Institute of Technology demonstrated that it was very unlikely that Theia should have the same composition as Earth. This led to some interesting adjustments to the prevailing model.

The Caltech team proposed that immediately after the impact, while Earth and the proto-Moon were molten, the whole system was enveloped in a vaporous cloud of material that became homogenized by convection, in effect stirring up the material and blending it as you would when combining fluids in a drink. The fact that Earth rocks and Moon rocks show identical isotopic signatures essentially forces this model to the fore, and it is a very realistic scenario. It does mean that the forming lunar disk, however, must have existed for at least a century to enable the degree of mixing needed to have taken place.

Still, there are some difficulties in explaining everything involved with the Theia concept. Planetary scientists would expect that a magma ocean on Earth would have resulted at first following the impact, and there is no evidence that it did. Ratios of volatile elements on the Moon are not explained well from this idea. It is still difficult to understand the isotopic ratios of oxygen and titanium being identical between Earth and Moon – could Theia also have had identical oxygen and titanium isotopes?

Most planetary scientists are nonetheless relatively confident about the modi-fied, mixing scenario, enabling the Giant Impact Hypothesis to work. The identical isotopic ratios could be explained if turbulent mixing followed the impact. The Moon's crust is rich in anorthosite, an igneous rock rich in plagio-clase feldspar, which you would expect if the Moon's surface was once molten. This widespread melting is further supported by the existence of KREEP-rich samples in Moon rocks, specimens that are rich in Rare Earth elements such as rubidium and lanthanum. (KREEP stands for K as in potassium, REE for Rare Earth Elements, and P for phosphorus.)

A variety of evidence – among them density, moment of inertia, magnetic induction, and rotational signature – suggest the Moon's iron core is small, less than 25 percent the lunar radius, or about half that of most terrestrial moons and planets. This suggests that the Moon indeed seems to have formed from the material of planetary mantles and not so much from cores, suggesting the core of Theia was absorbed mostly into Earth, which would also explain the angular momentum of the Earth–Moon system. This would have been the case because Earth would have gained significant angular momentum and mass from such a collision.

The metallic element zinc also provides some supportive evidence for the Giant Impact Hypothesis. Because the isotopic nature of zinc is heavily affected when it has vaporized, planetary scientists can straightforwardly tell when this has happened, as compared to normal geological processes. The Apollo lunar samples contain a substantial amount of heavy isotopes of zinc, but overall less total zinc than Earth rocks, which suggests a good deal of that element was lost in vaporization. This is consistent with what astronomers predict from the Giant Impact Hypothesis.

There seems to be some tenuous evidence supporting the odds for a giant impact from looking at other star systems, too. Astronomers have used the Spitzer Space Telescope to observe the dusty disk around the star HD 172555, a member of the Beta Pictoris Moving Group, some 95 light-years away. This star was first noted as extremely bright in the mid-infrared in the 1980s. In 2004, studies with Spitzer demonstrated that this star's infrared emission is far

Figure 4.5 Earth rises above the lunar horizon in this image made with the orbiting Japanese spacecraft Selene.
JAXA/NHC

brighter than would be expected from the star itself, which is about 12 million years old.

HD 172555 is a young, white-hot star twice as massive as the Sun and with a luminosity 9.5 times greater than our star. Based on our current ideas about planetary formation, astronomers believe this system is in the process of forming rocky planets. And, amazingly, Spitzer detected a huge amount of silicon-rich material, in the form of silicon dioxide gas and amorphous silica. These are in stark contrast to more "normal" forms of silicon on Earth, such as silicate minerals olivine and pyroxene.

In 2009, a group led by Carey Lisse of Johns Hopkins University examined the system carefully using Spitzer again, along with results from the Deep Impact and Stardust cometary missions. The results from this study were quite amazing. The scientists found a region of warm material about 6 astronomical units from the star, and consisting of huge amounts of vaporized rock along with substances such as frozen magma, lava, and rubble. The distance in this system is equivalent to the inner edge of the Main Asteroid Belt in our solar system.

The group's conclusion was that they had observed the products of a huge impact between two terrestrial planets, analogous to the Giant Impact Hypothesis in our own system. The velocity of this impact had to be very high, as well, in order to transform ordinary silicates (olivine, pyroxene, etc.) into a gas. The group also concluded that rocky protoplanets exist in this system.

The mounting evidence for the Giant Impact Hypothesis raises questions about that now-gone impacting body, too. If Theia existed, where did it come from? In 2004, Princeton University astrophysicist J. Richard Gott III (1947–) and his colleague Edward Belbruno (1951–) advocated the origin of Theia at one of the Lagrangian points, stable orbital positions far from Earth, proposing L_4 or L_5 as the likeliest places. These are positions that either precede Earth or follow it in its same orbit, and are offset by 60°. The Princeton scientists proposed that Theia's growing mass would have knocked it out of the stable point and into a collision course with Earth.

Other modifications have also been proposed in recent times to clean up some of the remaining difficulties with the Giant Impact Hypothesis. The timing of the impact might have been 4.48 billion years ago rather than the generally accepted value of 4.53 billion years. Theia may have smacked into Earth head-on, rather than simply sideswiping it. This could have happened if the velocity was much higher than originally thought, and the collisional speed could have essentially destroyed Theia. This allows for a more forgiving model of Theia's composition.

Robin Canup, the astronomer who reignited the Giant Impact Hypothesis momentum, has also put forth recent modifications to consider. In one alternative explanation, she suggests that perhaps Earth and the Moon were created from the collision of two bodies, each much larger than Mars, that accreted to form Earth. Part of the accretion process included a ring around Earth that formed the Moon.

In 2003, Canup produced a substantial paper for the scholarly journal *Icarus* providing the results of about 100 hydrodynamic simulations of potential Moon-forming impacts. This influential paper demonstrated that in such a Giant Impact, the Moon could have formed from a massive, iron-poor disk of material consisting of particles that were minimally heated during the collision, and therefore escaped being absorbed into Earth. These very sophisticated simulations suggest that Theia had a mass between 0.11 and 0.14 times that of Earth.

As we have seen, the origin of the Moon is one of the oldest problems of planetary science – going back in some form to some of the earliest thoughts humans had when looking up at the sky. It is a highly complex issue, and one that is anything but simple to resolve. We don't know exactly how the Moon formed, and not having been around four and a half billion years ago means that we have to reconstruct a terrifically complex detective story a long time after the fact. But the vast majority of evidence points strongly to the Giant Impact Hypothesis as the answer, one that we didn't have at all until just 40 years ago.

There is, however, a caveat to this story. It is not yet absolutely known to be true. Some planetary scientists feel the model of the Giant Impact Hypothesis has been stretched to great lengths to accommodate findings as time rolls on. Despite the consensus on the impact, planetary scientists don't yet really understand this event. The summary is not yet quite satisfying. It reads that large-body impacts in the early solar system were likely. In some scenarios, it is

Figure 4.6 Neil Armstrong pauses for a moment to take a snapshot of the far-off lunar module during Apollo 11.
NASA

possible to put iron-poor, silicate debris resulting from such a collision into orbit. This debris, if accreted into a single body, would have properties similar to the current Moon.

The plausibility of the Giant Impact Hypothesis is solid. Planetary scientists agree on that. But we do not yet undeniably know that it happened. We are inside the detective story of the origin of the Moon, and perhaps getting toward the mystery's solution. But for the time being, we have to be satisfied with being not quite there yet.

Chapter 5
Where has all the water gone?

Deciphering weather and global climate on another planet is not easy. As anyone knows from watching local news meteorologists, it is not easy right here on Earth. Mars, the Red Planet, is a special partner in the solar system, being a terrestrial world that is relatively close by and characterized by a desert-like climate, sandy dunes, wind storms, polar ice caps, and other features that make it seem similar to various locales on Earth.

But in reality, Mars is terrifically different than our world. With an equatorial diameter of 6,792 kilometers, the Red Planet is slightly more than half Earth's size, and it contains only a tenth of Earth's mass. The martian orbit carries it around the Sun once every 687 days, making its year equivalent to 1.9 Earth years.

Like Earth, Mars is differentiated – that is, it has a dense metallic core overlain by a rocky mantle. The planet's familiar orange color comes from copious amounts of iron oxide, like rust, richly coating the planet's rocks. Mars' two moons, Phobos and Deimos, were discovered by American astronomer Asaph Hall (1829–1907) at the US Naval Observatory in 1877; they are probably captured asteroids, and measure a mere 27 by 22 by 18 kilometers (Phobos) and 15 by 12 by 10 kilometers (Deimos). Alternatively, some planetary scientists believe they may be pieces of Mars from large impacts.

Over the years, spacecraft missions have added enormously to our understanding of Mars. Missions got off to a rocky start, however. Following a series of unsuccessful missions by the Soviet Union and the failed US probe Mariner 3, the first flyby of Mars took place when the Mariner 4 spacecraft flew past the Red Planet in 1965. Mariner 9 produced a flood of martian imaging in 1971–1972. Substantial progress commenced with the US Viking 1 and Viking 2 landers, which touched down on Mars in 1976 and conducted various experiments. More significantly than the landers, however, were the Viking orbiters, which conducted vast amounts of science after the landers proved somewhat disappointing.

Twenty years later, in 1997, the Americans landed the Mars Pathfinder spacecraft in Chryse Planitia, a smooth circular plain. The craft consisted of the Carl Sagan Memorial Station, named after the then recently deceased astronomer, and a small rover called Sojourner. One month earlier another American craft, Mars Global Surveyor, entered orbit about the Red Planet and operated until 2006.

Sadly, in 1999, two US missions were lost at Mars – Mars Climate Orbiter and Mars Polar Lander/Deep Space 2. In 2001, however, NASA's Mars Odyssey arrived in martian orbit and began science operations, which continue today. In late 2003, the European Space Agency's Mars Express mission arrived in orbit and deployed the Beagle 2 lander. Unfortunately, the lander was lost, but the orbiter continues to operate.

Next came a major milestone in Mars exploration. NASA's Mars Exploration Rovers, Spirit and Opportunity, landed on the Red Planet in 2004 at two widely separated locations – Spirit at Gusev Crater, a suspected former lake in a crater, and Opportunity in Meridiani Planum, a flat plain. Spirit became stuck in soft soil in 2009 and the following year lost communication with Earth, ending its mission. Opportunity continues to operate on Mars.

During the operation of the two rovers, NASA also launched the Mars Reconnaissance Orbiter (MRO), which arrived in martian orbit in 2006. This highly productive mission continues to operate. In 2008, NASA's Phoenix spacecraft landed on Mars and studied the geological history of martian water until the craft was essentially spent that same year. In 2010, the spacecraft was briefly revived, operating for a few more months. The latest major Mars mission, Mars Science Laboratory (MSL), landed in Gale Crater in 2012 and the craft's large, sophisticated Curiosity rover continues science activities. A NASA probe launched in 2013, Mars Atmosphere and Volatile EvolutioN (MAVEN) will also attack the questions surrounding martian water.

Make no mistake about it, the exploration of Mars is a red hot cottage industry because the planet shows evidence of abundant water in its history, and where there is water, there could also be life.

But astronomers don't all see exactly eye to eye on how much water existed on Mars and exactly when it was there. Many planetary scientists believe Mars had abundant surface water – an ocean – covering the lowlands on much of the martian northern hemisphere. The opposing view says that Mars had standing surface water for brief periods only, and perhaps in regional or local areas only. This view suggests that even during the times when large volumes of water existed on the surface and carved riverbeds (imaged by a variety of spacecraft), the freely flowing surface water was really temporary.

So while the verdict is out, most planetary scientists believe that during the first billion years of the Red Planet's history, liquid water existed on the surface, at least intermittently. They also hold that relatively low places hosted lakes of water that existed for at least short periods, and that lakes in crater floors also existed for appreciable amounts of time. Recent evidence shows that such lakes were present more recently than the first billion years.

The outstanding question of martian science is: What transformed Mars from a wet planet to a cold and dry one? A huge array of data addressing this question has bombarded us over the past decade. The answer isn't yet entirely clear, but tantalizing clues are mounting.

Planetary scientists believe Mars had a denser atmosphere early in its history. This may have helped to make the planet warmer and helped to keep water from evaporating into space as easily as it does now. Clearly, a climatic change or a series of changes have transformed Mars, ending the planet's active wet phase or making such phases much less frequent. We know from a variety of sources that ample water exists in ices (at the poles and mostly in the subsurface) and probably as a liquid underneath the frozen surface, in deep aquifers. The planet's weak gravity allowed much of the early water to escape, however. (The planet's greater distance from the Sun than Earth also means it is colder and molecular motions are slower.) Mars also lost much of its magnetic field, possibly due to an early impact, and this allowed the stripping away of the atmosphere by the solar wind. Asteroid impacts could have also blown water molecules into space. As Mars grew colder, the remaining water was absorbed into the surface and froze, creating subsurface ice deposits.

The evidence for ice on Mars is rich and varied. Circumstantial evidence abounds from various spacecraft, such as MRO imaging shallow craters of only about 1 to 3 meters depth with ice exposed at mid-latitudes. In 2008, the Phoenix lander detected ice exposed in the top several inches in the planet's north polar region. The martian polar caps themselves contain substantial amounts of ice. And impact craters show areas of muddy debris around their collars. These could have been formed from meteorite impacts that melted subsurface ice and splattered debris upward and outward.

How deep does ice exist below the martian surface? It likely extends several kilometers below the surface, and scientists believe that below the ice aquifers of liquid water exist. Certainly, the evidence for Mars' geologically recent volcanic activity means that subsurface heat has warmed ice to liquid temperatures. But no one yet knows how long liquid water can flow under the martian soil. Perhaps the aquifers are short lived and refreeze relatively quickly. Or maybe the water is so salty that it never freezes. And wouldn't you like to sample these aquifers or regions of subsurface ice to see if any microbial life exists there or once existed there? It is one of the ultimate goals of martian exploration, and we aren't there yet.

Just as with other complex topics on other worlds, the real interest over martian water, ice, and climate change exists in the details. There are plenty of them. And they begin with an interesting conundrum.

The observations spacecraft have made of past water-eroded valleys and of gullies that are more recent, apparently carved by liquid water on steep slopes, both defy the current martian conditions. Pure liquid water, without much salt to lower the freezing point, could exist on the surface of Mars when temperatures are greater than 32 °F (0 °C) and when the partial pressure of water is greater than 6.1 millibars. Temperatures can exceed 32 °F in the upper centimeter of martian soil, which means if the right pressure exists, liquid water could exist on Mars' surface. But these conditions are unlikely today on Mars; the pressure would not be so great, and liquid water is highly unstable (often boiling) on the surface and rain could not occur.

However, liquid water could exist and be stable *near* the surface of Mars. In most areas of the planet, water that seeps upward to the surface would boil from pressure and then freeze from temperature. And ice would sublimate directly to a gas. However, at low elevations, water would not boil but would stay on the surface, frozen. Water might exist on the surface for short periods, and this might especially be the case at higher latitudes because of accompanying ice in the soil. But scientists are not yet certain of this. And where salts are present in the soil, liquid brines could persist at temperatures much lower than 32 °F.

Part of what enables water to be stable at high martian latitudes relates to the so-called frost point temperature. The temperature at which frost forms varies with the exact local amounts of water content. If the atmosphere of Mars cools to the frost point temperature, around −100 °F (−75 °C), ice condenses. And if ice warms to this temperature, it sublimates. At high latitudes, the ground nearest the surface warms to higher than the frost point temperature and is without ice. From a few fractions of a meter to a kilometer or so down, ice remains stable. At lower latitudes, soil near the surface contains several percent of water.

Overall, temperatures on Mars are normally below freezing all over the planet, and so, typically, the ground is frozen. The planet's two permanent polar ice caps vary in size and shape seasonally, but always contain a substantial amount of ice. Wintertime at one of the poles means it is continuously shadowed and contains some 30 percent of the atmosphere in the form of mostly carbon dioxide ice. In sunlight, much of this ice sublimates into the atmosphere once again. The relatively thin layers of carbon dioxide are deposited over substantial amounts of water ice, which make up most of the polar caps. Each polar cap extends out to about 80° latitude and contains stacks of layered sediments, deposited to a depth of about 3 kilometers. The layers contain mostly ice but also dust. These sediments are relatively young, and undergo cyclic seasonal variations, with much of the north cap perhaps dating to 100,000 years – but, of course, the upper layers are much younger.

Planetary scientists use the term cryosphere to define the zone on Mars where, if water exists, it is in the form of ice. How thick the cryosphere is depends on a variety of factors, including temperature, conductivity, and salinity of the groundwater. Planetary scientists have estimated the thickness of the cryosphere to range from 2.3 kilometers at the equator to 6.5 kilometers at the poles. When the planet was warmer in the past, there may not have been a cryosphere at all. But the cryosphere now probably holds a great deal of water in the form of ice, despite the fact that such ice tries to migrate as vapor toward the surface where it can sublimate. Salty water may also exist within the cryosphere. And underneath the cryosphere lies the martian hydrosphere, the region where even pure water could be liquid. Exactly how much water exists there is not yet known.

Figuring out what happened to the water on Mars is especially urgent for planetary scientists because it is the common thread between the Red Planet and Earth. Various spacecraft have revealed long channels on Mars that extend hundreds or thousands of kilometers long, most significantly in the areas of the large basins Chryse, Elysium, and Hellas. These basins may have once held significant bodies of water.

Mars scientists call another type of water-related feature, branching valley networks, which occur mostly in the cratered highlands of the Noachian era. Gullies on Mars were also thought to be formed by running water, and appear on steep slopes in a variety of areas on the planet. These are ongoing features, actively forming today, as observed extensively by MRO. Both the Opportunity rover (at Meridiani Planum) and Mars Express found evaporate minerals that formed from water, and they found structures indicative of flowing and standing water on the planet.

Additionally, scientists have imaged so-called layered deposits in a variety of areas that may be related to past water on the planet, or to wind-blown deposits or volcanic ash. These deposits exist in a number of places such as inside Valles Marineris, the system of large canyons discovered in 1971 by the NASA spacecraft Mariner 9. One of the largest valley systems in the solar system, Valles Marineris stretches along the martian equator, east of the Tharsis Bulge. Thick, layered deposits exist in many chasma throughout the great rift, extending from canyon floors up to the heights of the canyon rims.

Figure 5.1 The long scar of Valles Marineris stretches 4,000 kilometers (2,500 miles) across the martian surface, 10 times the length of the Grand Canyon.
NASA/JPL

No one knows how these layered deposits formed. They definitely formed after the opening of Valles Marineris. Whether these sedimentary layers are remnants of weathered rocks or are young deposits that accumulated relatively recently is open to interpretation. A strong possibility, however, is that the sediments were deposited in huge lakes. Outflow channels also appear to originate from the eastern ends of the canyons, which would also support water flowing through these areas. And just in the last few years, spacecraft have discovered sulfate minerals in the canyons such as kieserite (magnesium sulfate hydrate, $MgSO_4 \bullet H_2O$), and gypsum (calcium sulfate hydrate, $CaSO_4 \bullet 2H_2O$). The leading idea is that these minerals formed from groundwater upwelling.

But there are problems with the idea of great lakes in this region. The canyons must have been closed at their eastern ends long ago. (They aren't now.) And how could the lakes have filled? If it was by rainfall, then the martian atmosphere must have been substantially warmer in the past. They also might have formed in a colder climate from the melting of substantial amounts of ice. The faults that caused the canyons themselves could have allowed water to seep upward. Or lakes could have formed for relatively short periods during a transition between a warmer and colder overall climate in a variety of ways.

Other types of terrain point even more strongly to the presence of substantial amounts of water on Mars's surface. Outflow channels and valleys betray strong evidence of abundant past water, and were first imaged by Mariner 9 in 1971. But acceptance of their watery origins came along later. These are some of the strongest pieces of evidence that point to a wet early martian climate.

In martian geography, the term outflow channel refers to a linear feature that can stretch for hundreds of kilometers and marks the former flow of liquid water across terrain that may still show some remnants of its original makeup. These channels formed during times of flooding. Valley refers to a feature typically a few kilometers across, but stretching over long distances. Valleys may have formed by slowly running water etching away soil on the surface. Gullies are small features on steep slopes that may be only tens of meters wide and less than a kilometer long.

The largest outflow channel on Mars is Kasei Vallis, which originates in Echus Chasma, near Valles Marineris, and empties into Chryse Planitia, after stretching for more than 2,400 kilometers. At times spanning 400 kilometers across, it reaches a maximum depth of about 2.5 kilometers. Many of these channels contain teardrop-shaped "islands" around which water presumably flowed. If these channels formed by floods, as is the prevailing view, then they suggest enormous quantities of water freely made their way across the martian surface. But they also present conundrums. The volume of water would be so great in some channels that it would have had to be stored, and then quickly released and dispersed. No evidence of large bodies of water at the "ends" of these channels exists.

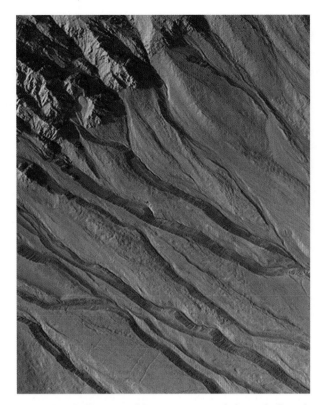

Figure 5.2 Gullies resembling water-carved channels flow down a crater wall on the Red Planet's southern highlands in this image captured by the High Resolution Imaging Science Experiment on NASA's Mars Reconnaissance Orbiter.
NASA/JPL/University of Arizona

Valley networks are quite common in the southern cratered highlands of Mars. First observed in 1972 during the Mariner 9 mission, they have provided ample controversy over their origins ever since. The great majority of planetary scientists agree the valleys were chiseled by running water. But what that means about past martian climate conditions is yet up in the air.

Most valleys were laid down in the Noachian era, 4.1 to 3.7 billion years ago, but recent finds indicate some of them formed during the Hesperian era, 3.7 billion to about 3 billion years ago, or perhaps even more recently. If valleys required warm climates in order to form, what does this say about more recent martian history?

Terra Cimmeria, a large region in the martian highlands, is heavily marked by valley networks, as is a large, broad region of the planet just south of the equator from 20° east longitude to 180° east longitude. Some valleys stretch as long as 1,000 kilometers, and many valleys are clustered together, show short tributaries, and typically are 50 to 200 meters deep. Images made with the Mars Global Surveyor's Mars Orbiter Laser Altimeter (MOLA) instrument show

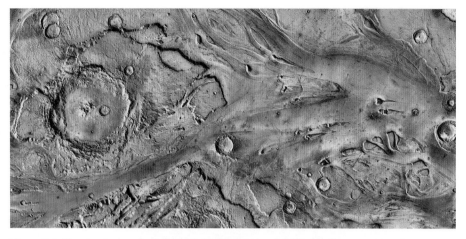

Figure 5.3 Floodwaters rushed through Valles Marineris billions of years ago and across Mars' northern lowlands, creating this channel known as Kasei Valles, shown here in false color. Red and yellow areas contain rock outcrops and harder sediments, whereas blue and green tones are home to sandy and dusty material.
NASA/JPL-Caltech/University of Arizona/HiRISE

spectacular views of valleys such as Samara Vallis, Loire Vallis, and Parana Valles, all in Terra Meridiani.

Gullies appear on the slopes of quite a few martian features such as crater rims, pits, and the walls of valleys, and begin as somewhat broad, irregular features that narrow sharply into linear or arcing channels that are meters or tens of meters wide and a few hundred meters long. They can also widen and form tree-like branching groups of channels. Most gullies are in the martian southern hemisphere at latitudes greater than 30°. The fact that so few appear in the planet's northern hemisphere is simply because the south has more features with steep slopes.

Generally speaking, planetary scientists do not feel gullies result from groundwater seepage. Another correct, proven hypothesis is that they are caused by dry material simply slumping downward, aided by carbon dioxide. Another possibility is erosion by water originating from melting ice. Because these features appear to have formed recently, water was a likely suspect, but the water's source was not clear.

More significant areas of water certainly existed on Mars in the distant past. The largest of these, lakes and oceans, must have filled basins of many different sizes. Terrain from the Noachian period is very much eroded. After the Noachian period, the evidence for large amounts of water on Mars mostly comes from the outflow channels. Periods of flooding probably resulted from colder conditions and a thick cryosphere, and estimates of the amount of martian water that could have made seas ranges from a global coating of a few hundred meters to 3 kilometers deep.

The evidence of depressions on Mars that once hosted so-called paleolakes comes in a variety of forms. Eroded remnants of deposits that look like a lake outflow exist at Parana Valles. Several depressions in the region of Terra Sirenum contain such deposits, and range up to 200 kilometers across and 1 kilometer deep. Quite a number of craters have rims that appear to have been breached by an outflowing stream. One example is Gusev Crater, the landing site for the Spirit rover. Other craters show differing traces of evidence of past lakes. Terraces around the peripheries of craters could be the remnants of floors now eroded away. Deltas are commonly found, as with Eberswalde crater and many others. These deltas mostly post-date the Noachian.

When it explored Gusev crater, the Spirit rover found little evidence of a previous lake. The rover explored a plain strewn with volcanic rocks that did not appear to be significantly eroded, and was without sediments. Southeast of the crater, when it explored the Columbia Hills, Spirit found some rocks

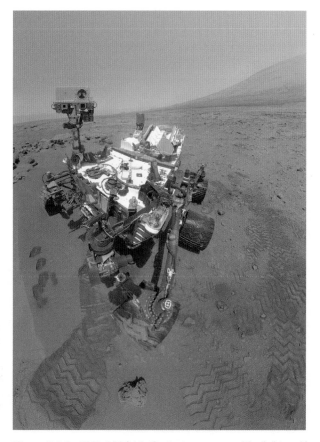

Figure 5.4 In 2012, NASA's Curiosity rover grabbed this self-portrait, a 55-image composite, after scooping four samples of ruddy dirt in Gale Crater. The vehicle is searching for environmental signs that microbial life might have once been possible on Mars. NASA/JPL-Caltech/Malin Space Science Systems

indicative of hydrothermal alteration. But planetary scientists feel certain that Ma'adim Vallis, which enters Gusev crater, was formed by flowing water and that Gusev must have held water long ago. The eroded materials may have been buried by later volcanic activity.

The big basins, Argyre and Hellas, are also interesting places to imagine ancient lakes. Argyre may have contained a large Noachian-era lake that flowed out to the north, as evidenced by disturbed terrain extending from the basin's north rim to Margaritifer Terra. Winding channels that lead from here to the Chryse basin, if they were flooded and ran continuously, would form the largest networked river complex known in the solar system. Apparent outflow channels from Argyre exist in Uzboi Vallis and Ladon Vallis, and from Hellas they exist in Surious Vallis, Dzigai Vallis, and Palacopus Vallis.

The large basin called Hellas contains the lowest point on the planet, and so certainly running water could have formed a large lake here. In 2001, scientists examining the imagery of Hellas hypothesized two former shorelines extending around the basin at two different elevations. It is possible that a significant lake in Hellas developed a kilometer-thick ice covering that eventually slumped downward, creating strange landforms in the basin that we now see.

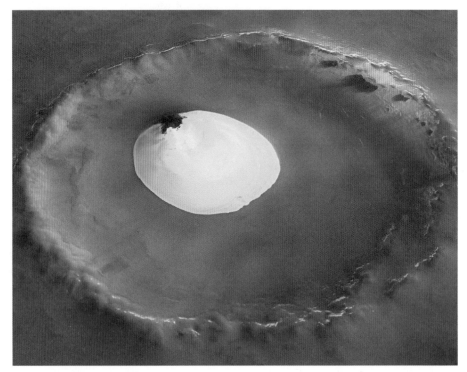

Figure 5.5 Water ice solidifies in many places across the martian surface. The European Space Agency's Mars Express orbiter captured this striking crater in Vastitas Borealis, a northern lowland, where the water ice remains year around.
ESA/DLR/FU-Berlin

The sparsely cratered plains of the martian northern hemisphere stand out starkly compared with the heavily cratered south. The cratered terrain in the south exists at a higher elevation, and the northern hemisphere plains are lower, and seem to have been prone to flooding in the distant past.

Two significant depressions exist in the north, Utopia basin and the North Polar basin. Utopia basin is the largest impact basin on the planet. Flooding in Chryse Planitia also undoubtedly filled the North Polar basin; Amazonis Planitia, the flattest region of the planet due to flood lavas, would also have flooded.

The evidence of past shorelines is also controversial among the northern plains of Mars. Several indicators of this terrain include cliffs that separate plains from highlands, plains with winding ridges, valley walls with terraces, "island" terrain within plains that shows sloping, and small outflow channels. Hypothesized shorelines have been imaged and studied in the Deuteronilus Mensae, along the North Polar basin, south of Elysium Planitia, around Meridiani, and in other locales. Some ocean proponents argue that stable shorelines would never have formed due to constantly changing sea levels.

The fate of a northern ocean, if it existed, has also been the focus of scientific work. Such an ocean would have accumulated in a connective stage, frozen as the planet cooled, and eventually sublimated.

The hunt for signs of water on Mars has accelerated over the past 15 years as rovers took to the ground in several key spots. When it landed in 1997 in Chryse Planitia, Mars Pathfinder and its Sojourner rover were a glorified engineering experiment sent in part as a test to see how experiments on other planets could work. The rover could travel only a short distance and carried an x-ray spectrometer to analyze rocks and soils. The area where Pathfinder landed is thought to be a depositional fan of Ares Vallis, a flood channel entering Chryse from the south. The surface topography contained rocks ranging from boulders down to pebbles. Some of the rocks are oriented as if they were pushed along by a channel of water that flowed through the area. Many rocks are apparently eroded by small particles, and sand litters parts of the area.

In 2003, along came Spirit and Opportunity, and they explored very different areas. Mission planners believed both areas would show evidence of liquid water interactions. Each rover had a miniature thermal emission spectrometer (Mini-TES) as well as a movable arm with several instruments, including two spectrometers and a rock abrasion tool (RAT) for grinding samples. Spirit landed in Gusev crater and Opportunity in Meridiani Planum.

Spirit landed where, as we have seen, water must have pooled in ancient times. The rover did not identify lake-related sediments, but did find hills within the crater that clearly had been acted on by hydrothermal water. Spirit came down near the center of Gusev and after studying the landing area, made its way to Bonneville Crater and then to the Columbia Hills. Some of the Columbia rocks studied, those of the so-called Peace class, contained a magnesium-calcium sulfate component, strongly suggesting they were deposited in a watery environment.

Figure 5.6 NASA's Opportunity rover entered Endurance Crater in 2004, where it examined layers of bedrock, sediments, and sand dunes. While climbing out of the crater, Opportunity also discovered the first meteorite ever found on another planet.
NASA/JPL/Cornell University

Opportunity came down on a level plain relatively free of craters. This area was chosen because of the orbital detection of large amounts of hematite (iron oxide, Fe_2O_3), which could have formed amid water. The spacecraft landed inside the shallow Eagle crater. The lander imaged numerous tiny, spherical pebbles nicknamed "blueberries" about 5 millimeters across that contain abundant hematite. Unlike those in the Columbia Hills, the rocks here seem to have been influenced by water. The rover then explored Endurance crater, where it studied a curious cliff called the Burns formation. There it analyzed many minerals, including silicate hydrates, and many sulfates, chlorides, and phosphates, all of which require ample water to form.

On examining Burns cliff carefully, scientists found it holds compelling evidence of abundant water. Rectangular voids speckled through the rock outcrops mark the locations of soluble crystals that dissolved away after the rock had hardened. The hematite-rich blueberries probably formed in less than 1,000 years following the decomposition of jarosite (potassium iron sulfate hydroxide, $KFe^{3+}_3(SO_4)_2(OH)_6$) or oxidation of iron-rich sulfates.

Opportunity scientists believe they identified at least four episodes of groundwater moving in and out of the area of Burns cliff. The fact that jarosite was discovered in this region means the water must have been rich in sulfates and very acidic. The record of the rocks gives away the episodes of water flowing in and out, and suggests water helped to create the hematite-rich blueberries. Opportunity's analyses at Meridiani suggest that abundant brines evaporated, the brines having formed from acidic, sulfate-rich groundwater interacting with basaltic rocks.

The Curiosity rover, part of the Mars Science Laboratory mission, landed in Gale Crater in 2012 and continues its mission there. Again, of course, Gale was chosen as a good site in which to investigate a watery past. The rover, much larger than its predecessors, carries a suite of spectrometers, a rock abrasion tool, and many other instruments. The area called Yellowknife Bay explored by Curiosity seems to be an ancient lakebed in a defunct river system, and appears not to have been as oxidizing or acidic as other martian environments visited before, along with a different array of salts. Areas of stream-worn pebbles have been studied by Curiosity. The analytic labs definitely measured clay minerals.

So the evidence of abundant water flowing on the surface of Mars, in its ancient history, is compelling. Although arguments have been pushed back and forth over the past 2 decades over how strongly the evidence supports an early, wet Mars, the evidence now is strong enough such that most planetary scientists assuredly believe in it. The Noachian era includes periods during which lakes formed, although other periods of the Noachian were dry. This notion is supported by a flood of data, including high erosion rates during the period, Meridiani's water-lain sediments, and the deposition of water-soluble minerals in Noachian rock sediments.

This evidence, of course, leads to our big question: What happened to Mars to transform it into a cold, dry climate?

In order for Mars to have been warm in the Noachian, the planet must have had an impossibly thick atmosphere that was actively warmed by a substantial greenhouse gas cycle. Planetary scientists believe the Sun, 3.8 billion years ago, was only 75 percent as luminous as it is now. Without a greenhouse warming effect, Mars' surface temperature would then have been only –107 °F (–77 °C), far colder than the 32 °F (0 °C) or more needed for pure liquid water. So presumably, the martian surface at that time would need to be atmospherically warmed by at least 139 °F (59 °C), which means it captured 85 percent of the radiation from its surface. (Earth's present atmosphere captures 56 percent of our planet's surface radiation.) Atmospheric scientists studying Mars have not been able to explain this. Furthermore, the geological evidence doesn't require this. It may have snowed but never rained, except after large impacts.

Another unusual possible explanation came from research in 2002, when astronomers proposed that perhaps early Mars warmed due to large amounts of rock and water vapor being infused into the atmosphere by large impacts. Molten rock at a temperature of about 2,500 °F (1,400 °C) could have rained down from the atmosphere, heating the planet overall, causing the evaporation of groundwater and ice. This would produce substantial rain events and cause a cycle that could produce valley networks.

Other problems confront planetary scientists trying to understand early Mars. Even with a thick atmosphere and a relatively warm martian surface, it is difficult to understand how the planet could retain a dense carbon dioxide atmosphere over long timescales. Impacts by asteroids and weathering would tend to reduce the atmosphere over time. Planetary scientists examining the number of Noachian era impacts suggest the planet may have lost 50 to 90 percent of its atmosphere during the period 4.4 to 4.0 billion years ago. Perhaps atmospheric stripping by the solar wind played a major role.

And weathering would also reduce the atmosphere's carbon dioxide content. It is possible that carbon dioxide could be exchanged in and out of the atmosphere, however, by the formation of carbonate minerals, and their return of carbon dioxide to the atmosphere. Volcanism could also have played a significant role in burying weathering products and returning carbon dioxide into the atmosphere.

Following the Noachian era, Mars changed dramatically according to some researchers. The rate of formation of water-driven valley networks declined substantially. The idea that Mars suddenly dried up and became much colder is borne out by a variety of evidence, including evidence of water seepages, little evidence of weathering on post-Noachian era rocks, eruptions of groundwater that required a thick cryosphere, and a steep decline in erosion rates.

Some planetary scientists have proposed that post-Noachian floods caused temporary oceans on Mars and episodic bursts of carbon dioxide into the atmosphere. They believe that Mars was cold and dry for much of its history following the Noachian, but that periods of 1,000 or 10,000 years saw brief returns to a warmer, wetter climate. But this idea doesn't fare any better than the greenhouse cycle models in trying to explain how the carbon dioxide disappeared. It is possible that the flooding that created the lakes and seas did not change the atmosphere for long. The standing bodies of water may simply have frozen in place, later sublimating and transporting to the polar ice caps.

Episodic warming also could have occurred due to impacts or volcanoes. And variations in the planet's orbit and rotation also must have affected its climate significantly. The planet's obliquity (the angle between the planet's orbital axis and rotational axis) and eccentricity (the amount its orbit deviates from a circle), could have far-ranging effects. In fact, the planet is now emerging from a period of high obliquity, greater than the present 25°, during which there may have been a deep martian "ice age." High obliquities can affect ground ice, and may have created an icy veneer over the planet, particularly between latitudes 30° and 60°.

All the answers about Mars' global history are not yet here. We do know that early Mars was wet and, post Noachian, underwent a transformation to a colder, drier climate. The reasons why Mars was wet – how it could have been – remain somewhat mysterious. A carbon dioxide-rich greenhouse effect seems to be inadequate, by itself, to produce such an early Mars. The post-Noachian era was mostly cold and dry but large floods occurred. They may have episodically changed the global climate, forming a number of valley networks and temporarily witnessing water flowing once again.

The ice, now mostly underneath the martian surface, along with presumed abundant aquifers of liquid water, holds the key to understanding the missing pieces of the story of Mars. Any life that could exist in these regions will no doubt be the ultimate target of future exploration of the Red Planet.

Chapter 6
Why did Venus turn inside-out?

Venus is unmistakable in our skies. Never straying terribly far from the Sun, it blazes brilliantly in the evening and morning skies, shining as brightly as magnitude –4.6, the most luminous permanent object after the Sun and Moon. Long called Earth's "sister planet," the similarity is slight. It lies relatively close to us in the solar system, the next planet inward toward the Sun. Like Earth, it is a terrestrial planet, a predominantly rocky body, is about the same size as Earth, just about 95 percent the diameter of our planet, and contains about 82 percent of Earth's mass. But that's where the similarity ends. In most respects, Venus could hardly be any more different than Earth.

On this world that lies about 30 percent closer to the Sun than Earth, ironies abound. They begin even with the planet's name, which comes from the Roman goddess of love and beauty. When this bright and mesmerizing light wandering among the stars was named, the idea might have made sense. But early spacecraft studies of Venus betrayed the hellish nature of the planet.

Venus played a key role in the turning point of understanding the solar system in 1609, when Galileo Galilei observed it with his newly constructed telescope, from Padua, Italy. Galileo watched the planet and sketched it over the course of months, seeing that it underwent phases analogously to the Moon.

But the nature of Venus itself remained largely mysterious until the first spacecraft missions visited the planet in the 1960s. The first such probe, the Soviet Venera 1, was launched in early 1961. Mission successes were slow, from both the Soviet Union and the United States, marked by spacecraft failures. In 1962, the US craft Mariner 2 became the first successful interplanetary mission, measuring the surface temperature of Venus to be a searing 425 °C (800 °F), and ending speculation that the planet might harbor life.

In 1966, the Soviet Venera 3 probe became the first spacecraft to enter the atmosphere and crash land on the surface of another planet. A year later, Venera 4 measured the planet's surface to be even hotter than Mariner 2's result, and showed the planet's atmosphere consists of 90 to 95 percent carbon

BEHEPA-9   22.10.1975   ОБРАБОТКА ИППИ АН СССР   28.2.1976

BEHEPA-10   25.10.1975   ОБРАБОТКА ИППИ АН СССР   28.2.1976

Figure 6.1 In 1975, the Soviet Venera 9 spacecraft became the first to orbit Venus and return images from its surface. Several days later Venera 10 landed some 1,500 kilometers (900 miles) away from the Venera 9 site, and also returned images (bottom).
Soviet Planetary Exploration Program/NSSDC

dioxide. In 1969, Venera 5 and Venera 6 plunged deeper into the venusian atmosphere, collecting far more data, before being crushed by the planet's intensely high pressure.

The 1970s witnessed a continued, aggressive Venus exploration program from both the Soviet Union and the United States, with periods of significant cooperation. Soviet designers planned Venera 7, in 1970, to land and return surface data; its parachute probably was torn and the probe landed, compromised, able to return temperature data for some 23 minutes before failing. In 1974, Mariner 10 passed Venus at a distance of 5,790 kilometers and returned more than 4,000 images during its flyby. In 1975, Venera 9 and Venera 10 became the first probes to return images from the surface of Venus.

The US Pioneer Venus mission took place in 1978, with two craft – an orbiter that studied the planet for 13 years, and a multiprobe that entered the planet's atmosphere and studied it intensively. The Soviet program continued in high gear, with modest results from Venera 11 and Venera 12 and more impressive scientific results from Venera 13 and Venera 14. The latter lander returned the first color images of the planet's surface in 1982. The two last Venera probes, 15 and 16, orbited the planet in 1983 to produce radar maps of the venusian surface. The mid-1980s saw another burst of research on Venus when spacecraft set to study Halley's Comet, Vega 1 and 2, were also used – prior to their cometary encounters – to deploy descent craft to land on Venus and also drop a balloon package into the planet's atmosphere. The balloons studied conditions in the planet's atmosphere for about 2 days. Vega 2's lander was most successful, transmitting data for 56 minutes, and the balloon experiments worked.

Figure 6.2 The lava-shaped face of Venus hiding beneath the planet's thick atmosphere is revealed in this global map made with data from NASA's Magellan spacecraft.
NASA

This enormous interest in our "sister planet" in the early days of spacecraft exploration led to some amazing conclusions. But the Earth-shattering – umm, make that Venus-shattering – revelations would come from the Magellan mission, launched by the United States in 1989. The push for a radar-mapping mission of Venus began in the 1970s. US planetary scientists originally pushed for a mission called Venus Orbiter Imaging Radar (VOIR), but that concept proved too expensive for Congress. The scaled-down Venus Radar Mapper, ultimately renamed Magellan, operated for 4 years beginning in 1990, and provided vast knowledge of the geology of Venus.

Magellan produced many surprising and amazing results, but the biggest one came when the overall surface maps of the planet were fully realized. Venus has very few impact craters. This is an extremely strange result for a terrestrial planet in the inner solar system. Planetary scientists know that during the period of the Late Heavy Bombardment, some 4 billion years ago, numerous planetesimals and small bodies were flying around the region, slamming into bodies, accreting into larger objects, and generally wreaking havoc. All we need to do to see ample evidence of this is to look at the Moon and Mercury, for example. Earth has bountiful resurfacing mechanisms that have wiped those craters away. But what about Venus? How could a planet like Venus lose its impact craters? The result was certainly surprising, and you could even say that it almost made no sense.

Venus shows an amazing relationship between its atmosphere and its surface. The dense, carbon dioxide-rich atmosphere, very high temperatures, and pressure that is 90 times greater than Earth's, all give the planet a unique and hellish characteristic among terrestrial planets. The planet's atmosphere, consisting of its volatiles, says a lot about the venusian past. These volatiles form a link between the planet's interior, its surface, and its atmosphere, and they can change characteristics on timescales of millions of years. Convection plays a major role in the lives of terrestrial planets, as heat from a planet's interior dissipates. On Earth, plate tectonics, aided by ample amounts of water, reduces the strength of surface plates of rock, which move around, sink by subduction, and carry water back into Earth's interior. Plate tectonics has been a major process on Earth for at least 3 billion years.

But on Venus, no such thing as plate tectonics exists, although it may have operated in the past. Because of this, none of the major Earth features sculpted by plate tectonics – continents, ocean basins, systems of ridges at the boundaries of plates – exists on Venus. The majority of planetary scientists believe that plate tectonics never developed on Venus – or could have existed only briefly, very early on – because of the relatively low amount of water on the planet. If all of the water in the atmosphere of Venus were condensed into a layer on the planet's surface, it would amount to a layer only about 10 centimeters deep. And the ratio of deuterium to hydrogen in the atmosphere suggests that a huge amount of water that existed in the early history of Venus was lost into space.

Magellan's incredible discovery that Venus's surface is very young, relatively speaking, is the key to investigating the planet's geology, its history, and its future. Magellan captured synthetic aperture radar images and altimetry of the planet's surface over that 4-year range, recording more than 98 percent of the venusian surface. The craft also captured high-caliber measurements of Venus's gravity field. This mass influx of data on Venus gave planetary scientists a global model of Venus for the first time, and a huge amount of data to mine for years afterward, enabling them to propose coherent models for the planet's geological history. Magellan provided radar data down to about 120-meter resolution and altimetry data as precise as 1 to 10 kilometers. The business of understanding Venus began in earnest in the early 1990s.

Another spacecraft mission, Venus Express, followed more than a decade later. A product of the European Space Agency, Venus Express launched in 2005 and arrived in orbit a year later, beginning an intense period of study that continues today. The craft carries an instrument designed to study the interaction of the solar wind with Venus, a wide-angle CCD camera, a magnetometer, three spectrometers, and a radio sounding experiment. The many findings of Venus Express have included evidence of past oceans, an abundance of lightning, a south polar atmospheric vortex, an ozone layer in the planet's atmosphere, and a very cold atmospheric layer that may host dry ice.

Despite the current lack of plate tectonics on Venus, the planet has plenty of tectonic features and a wide-ranging array of volcanic activity. Shield volcanoes and large areas of lava plains similar to some found on Earth frequent the surface of Venus. Many of the highlands on the planet are formed around volcanic hot spots similar to upwelling volcanic areas on Earth such as those that produced Hawaii and the Canary Islands. Many unusual tectonic features on Venus, on the other hand, are unique to this planet. They include tesserae, multiple fractured zones of overlaying, crumpled terrain; and coronae, circular features believed to have formed from upwelling plumes of hot material.

When scientists first exhaustively made their way through the Magellan data, they were simply awestruck. Venus is a planet with a young surface, heavily volcanic in nature. Sometime in the past, major volcanic flows covered up the ancient surface that once existed. Researchers were struck with a mesmerizing question: What could have caused such a cataclysmic, relatively recent global resurfacing? What was it that caused Venus to turn itself inside out? (As one planetary scientist put it, "We are in the unenviable place of having to explain a planet that inexplicably threw up all over itself!")

Venus has about 940 named craters ranging between 1.5 and 269 kilometers in diameter. Because the planet's atmosphere is so dense, very small impacting

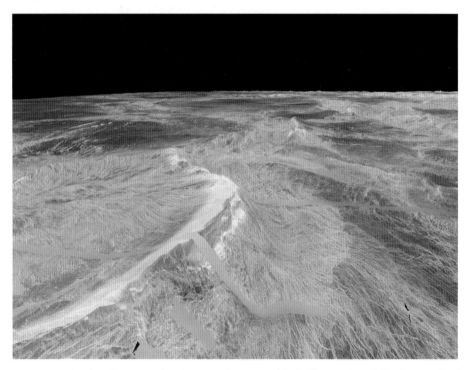

Figure 6.3 The deep lava troughs of Latona Corona and Dali Chasma stretch for thousands of kilometers across the surface of Venus, ultimately connecting volcanoes to highlands. This computer-generated image was made by placing radar data from the Magellan spacecraft over topography.
NASA/JPL

bodies of 1 kilometer or less generally disintegrate before they hit the surface, meaning that there are relatively few small craters on the planet. The explosion of small impactors in Venus's atmosphere and subsequent rain of small particles onto the surface may have caused some of the nearly 400 splotchy regions that appear either bright or dark in radar images.

Because no material from Venus has been analyzed in a laboratory, planetary scientists use the next best standard method of dating the planet's surface – counting craters. Planetary scientists believe they know the populations of small bodies in the solar system, at different times in the solar system, quite well. And dating samples of Moon rock and tying in the record of cratering on the Moon into those data gives them some ability to extrapolate this idea to the surfaces of other bodies in the inner solar system. But the different environments and histories and surface characteristics of other bodies like Venus mean that huge uncertainties creep into the picture.

Of course, surfaces of different planets behave differently, too. Erosion, tectonic activity, and other effects can eliminate craters. Taking all of this into account, the age of Earth's surface is very recent, because of the continuous new crust forming from plate tectonics and erosion. Earth's continental crust can be as old as 4 billion years; the oldest Earth rocks discovered include 3.962 billion-year old surface rock from the Northwest Territory in northern Canada and even older rock from crustal deposits in Western Australia, datable from the grains of zircon they contain, which in turn hold small amounts of radioactive elements. But on Earth, craters are wiped out quickly by erosion from water and wind. (Winds are a factor on Venus, but to a lesser degree than on Earth.)

By contrast, the ages of the Moon's surface and of the martian surface are known to be quite old, in the realm of 3 to 4 billion years. Not only have experiments on Moon rocks and on the surface of Mars borne this out, but analysis of ample numbers of lunar and martian meteorites recovered on Earth have also confirmed this result. Of 60,000 meteorites discovered on Earth, 194 are known to have originated from the Moon and 124 are known to have come from Mars.

Several decades ago, just the blink of an eye in scientific terms, we would have expected Venus to also have a surface that is 3 to 4 billion years old. But the weight of evidence from Magellan and other data – primarily from the cratering record – suggests a surface age for this planet more along the lines of 750 million years, which was shocking when it first came to light. Because of uncertainties in the analysis of venusian geology, planetary scientists accept a wide range for the age of between 300 million and 1 billion years. But the 750 million year figure is probably reasonably accurate. So how exactly did we arrive at this idea, and what exactly is going on with the planet's young surface?

To get started on the problem, scientists did what they always do at first – classify regions and types of features on the planet's surface. They did this prior to Magellan, but the mission in the early 1990s helped enormously to fix a more

Figure 6.4 On its way to Mercury in 2007, NASA's MESSENGER spacecraft snapped images of Venus' cloud-wrapped globe.
NASA/Johns Hopkins University Applied Physics Laboratory/Carnegie Institution of Washington

sophisticated understanding of the venusian surface. The most significant kinds of features on Venus are plains, tesserae, ridge belts, coronae, volcanoes, and mountain belts. They are not randomly distributed, but concentrated on parts of the planet. Magellan not only revealed those land features in greater clarity than before, but also uncovered new forms – wind streaks like those on Mars, outflow regions associated with impact craters, lobes of lava flows and channels, dunes, and impact craters as small as 3 kilometers across. The planet is dominated by low elevation volcanic plains, the huge solidified pools of lava from ancient eruptions. Some 85 percent of the planet's surface consists of volcanic plains, the result of several thousand volcanic vents, and 15 percent highlands stressed by tectonic activity.

The planet's highland features are notable, and appear spectacular in Magellan radar maps. An enormous southern highland, Lada Terra, is contrasted with smaller highlands such as Alpha Regio, Bell Regio, Eistla Regio, and Tellus Regio. Deep rifts connect some of these highland structures, and mountain and ridge belts as well as coronae accompany the apparent connections between some of these areas. Stress has deformed some of these areas, causing mountain and ridge belts.

The strange features called tesserae, previously termed parquet terrain or complex ridged terrain, appear nowhere else in the solar system other than Venus. Consisting of a complex network of ridges and depressions, these areas are marked by elevated plateaus of rough ground. The Magellan data tripled

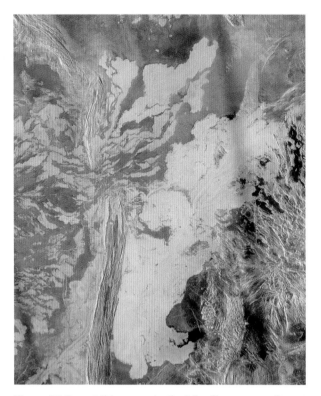

Figure 6.5 In a striking mosaic, the Magellan spacecraft captured lava flows in Ammavaru caldera. Radar bright and dark lava flowed for hundreds of kilometers before pooling in a vast radar-bright deposit at the center of this image, which covers tens of thousands of square kilometers on the planet's surface.
NASA/JPL

the number of known tesserae, and some of the best known are Ovda, Thetis, Phoebe, Beta, and Asteria. Between Ovda Regio and Aphrodite Terra stands the largest of these areas, comprising some 10 million square kilometers.

Those great depressed lava plains, making up the majority of the planet's surface, consist of a few different styles. They are ridge-and-band plains, band-and-ring plains, dome-and-butte plains, smooth plains, and patchy rolling plains. Studying these plains suggests stresses occurring on the different areas of Venus and happening, in terms of vertical and horizontal force, with different strengths in different areas.

The Magellan spacecraft opened a window on Venus by showing more than 1,600 volcanic landforms in unprecedented detail. And many features were shown for the first time, including coronae, deformed belts, and other landforms. Coronae appear more abundantly in the rolling plains rather than flood plains, but large amounts of lava released when the planet resurfaced 750 million years ago may have covered huge numbers of coronae in the lowlands.

Many unusual landforms and areas of interest occur on the planet. South of Ishtar Terra, which contains the four principal mountain ranges of Venus, lie extensive lowlands – Sedna Planitia and Guinevere Planitia. Ishtar Terra, larger than Australia but smaller than the United States, contains the impressive mountain chain Maxwell Montes, which stretches 11 kilometers high at its peak – taller than Mt. Everest. Sedna and Guinevere support the highland regions Alpha Regio and Bell Regio, as well as Eistla Regio, which contains the mountain peaks Sif Montes and Gula Montes, some 3.5 and 4.4 kilometers tall.

A belt of equatorial highlands stretches eastward from Guinevere, forming Aphrodite Terra, the largest region of highlands on the planet, rougher than Ishtar and about the size of Africa. Compression forces are suggested to be at play here, as the surface of this highland appears to have been buckled and cracked. Several massifs rise from this region, including Hestia Rupes, Ovda Regio, and Thetis Regio.

North of the area of Aphrodite lie the plains of Leda Planitia, Niobe Planitia, and Rusalka Planitia. South of Aphrodite are the plains of Aino Planitia. Large volcanic areas lie to the east, and Rhea Montes and Theia Montes, in Beta Regio, appear to be collapsed volcanoes resting in decrepit, worn-down calderas. The strange highland region Lada Terra is bordered to the north by Helen Planitia, Lavinia Planitia, and Aino Planitia. It is connected to Alpha Regio, to the north, by a network of rift valleys.

The venusian highlands are extremely unusual, fascinating areas. But they are secondary to the story of what happened to produce the extraordinary flow of lava, creating the vast lowlands. To understand the strange quandary they faced, planetary scientists knee deep in Magellan data turned to examining the lava-filled plains with special gusto.

The low areas on Venus show essentially two main types – lowlands and rolling plains. The unusual rolling plains typically measure from the average surface elevation – the venusian equivalent of "sea level" – to 2 kilometers below that spot. The lowlands simply are below the average surface elevation. Early on, scientists knew that the main lowlands made an unusual X-shaped formation with interference on their eastern side from Aphrodite Terra. North of this area, astronomers also recognized an unusual area showing the greatest spread in elevations on the planet.

Because they are relatively dark and are relatively smooth, scientists think of venusian lowlands as analogous to maria on the Moon. They are also relatively poor in impact craters, but do have some craters, areas of past flows of volcanic lava, volcanic domes and shields, and ridge belts. The deepest area of the venusian surface, Atalanta Planitia, is a centralized lowland flooded by volcanic lava and is one of the most important areas of the planet to study to understand how the basins, subsequently flooded with lava, formed. A ridge belt that crosses Atalanta Planitia is the most extensive on the planet, and these ridges offer evidence of the planet deforming on regional and smaller scales. The western part of Atalanta Planitia, however, contains ample dome-like hills.

As you might guess, the understanding of venusian plains underwent a renaissance with the first analyses of the Magellan data in the early 1990s. Magellan scientists identified smooth plains, which are pretty featureless; reticulate plains, which contain linear features spaced less than 5 kilometers apart; gridded plains, with sets of intersecting linear features; and lobate plains, showing lobes formed by flows at different times. Most of the types of plains contain wrinkle ridges, which are believed to have been formed by compression when forces squeeze areas of the planet at different rates. The production of these volcanic plains is impressive: Magellan identified 156 volcanoes more than 100 kilometers across, 274 volcanoes in the range of 20 to 100 kilometers, 86 caldera-like forms 60 to 80 kilometers across, 550 clusters of small volcanoes, and numerous smaller volcanic features like coronae, yielding more than 1,600 sources of volcanic lava eruptions on the planet. Most are thought to be long extinct.

Volcanism on Venus is unlike volcanism on Earth, on Mars, or on the Moon. On Earth, plate tectonics drives volcanism along the divergent and convergent plates, boundaries. And in areas associated with plates, hot spots – Hawaii, Iceland, the Galapagos Islands – are the recipients of volcanic upwelling. On the Moon, volcanism happened through the upwelling of extensive basaltic lava collecting on the floors of ancient impact basins. On Mars, volcanism became central, leading to huge shield volcanoes. On Venus, the atmospheric pressure is extreme, especially in the lowlands, and so volcanic outbursts would rise only a fraction as high as on Earth and pyroclastic flows (consisting of rock fragments and lava from volcanic eruptions) would be far less likely, despite the high temperatures.

Magellan's radar mapping of Venus produced a full survey of volcanic forms. Radar dark areas are relatively smooth, and radar bright areas rougher in texture. Analyses of large, radar-dark plains on Venus suggest they contain extensive fields of smooth lava like *pahoehoe* on Earth, as opposed to the rougher *aa*. These are very old flows, which have been eroded by wind, or worked on by the very reactive venusian air. Somewhat younger volcanic flows on Venus are both radar dark and radar bright, and show rough terrain from a large range of possibilities, including having oozed over rough terrain to begin with.

Some channels of lava flow on Venus have been identified to stretch as long as 250 kilometers. And many of the fields of lava are enormous, by Earth standards. In southern Atla Regio, south of the volcano Ozza Mons, a family of flows stretches 1,000 kilometers long and covers an area of some 180,000 square kilometers. Flows associated with Mylitta Fluctus, in the southern area of Lavinia Planitia, cover an area approaching 300,000 square kilometers. Studying such enormous flows, planetary scientists have estimated the rates of volcanic eruption for such large flows on Venus to be in the range of 460 to 4,600 cubic meters per second. This agrees with the rates of flow for some volcanoes on Mars, which are believed to have erupted from a centralized source. At Mylitta Fluctus, the original outpouring of lava is believed to have taken place over a span of 10 to 70 years, with smaller individual eruptions happening for a few days or at most a few months at a time since.

Data from the Venus Express spacecraft shook things up in 2010 when astronomers announced the discovery of three "hot spot" regions that have erupted on the planet's surface in recent venusian history. These areas, analogous to Hawaii on Earth, suggest mantle plumes underneath the venusian surface and appear to show recent outflows, as well as betraying differences in lava types at the three locations. The scientists suggest the flows are between 250,000 and 2.5 million years old.

More than 50 fields of major lava flows exist on Venus, some comparable to big ones on Earth such as the Columbia River Basalt Group in the northwestern United States or the Deccan Flats or Traps in west-central India. In the case of the Columbia River group, huge lava outflows resulted from long fissures or from long vents alongside fissures, and in the case of Columbia, the fissure system stretches 175 kilometers long. On Venus, there are subsurface dykes that extend hundreds of kilometers long, and these intrusions of magma were probably the sources of huge amounts of lava upwelling from below. Stresses inside the planet could have formed fissures through which enormous bursts of lava swelled upward, recoating much of the planet's surface in that huge event some 750 million years ago, largely resurfacing the planet.

The nature of the lavas on Venus is believed to be basalt-like, consisting mostly of amphibole and pyroxene, with plagioclase, feldspathoids, and olivine, and with some 20 percent quartz and 10 percent feldspar. Some venusian lavas may be more like a komatiite composition, with low content of silicon, aluminum, and potassium, and higher content of magnesium. Large outpourings of these lavas seem to have filled lowlands on Venus, resculpting the eastern region of Lavinia Planitia and Alpha Regio, in Sedna Planitia, in Beta and Phoebe Regiones, and on the surfaces of Aphrodite Terra and Atla Regio.

The smaller volcanic structures on Venus also have a tale to tell, and they include cones, domes, and shields. Tens of thousands of small shields exist, with most lying between 2 and 8 kilometers across and something like 200 meters high. They usually occur in clusters called shield fields. These shield fields are familiar to geologists. Earth has them, and they usually have central depressions that are linked to lava tubes, which feed the upward flow of lava but can also allow magma to channel back down, forming a pit.

The venusian lava cones are small, generally less than 15 kilometers across, and are also similar to Earth structures – volcanic cinder cones. They have steep slopes, and scientists believe they were built up from the upwelling of lava in the same way that cinder cones on Earth were.

On Venus, volcanic domes are larger than cones, some reaching 30 kilometers across and standing a kilometer high above the planet's surface. Many have pits placed at or near their centers. They are believed to have formed from the upwelling of highly viscous lavas such as dacites or rhyolites, which are thicker than basalt. They show sides and surrounding terrain that is radar bright – rough in texture – thought to have resulted from the chaotic, broken, violent effects this thick lava had on the domes and the surrounding landscape.

Some volcanic domes on Venus are steeply formed and show sharp scallops around their circular edges – that is, systems of radial lines coming from the dome's edges and forming a ray-like pattern around the domes. About 80 of these well-scalloped domes are known, and scientists believe they formed from the collapse of the dome's slopes, stressing the surrounding terrain and scattering material in the scalloped pattern.

In the wake of Magellan's mapping, many larger structures were systematically catalogued. The initial tally resulted in identifying 156 large volcanoes, 274 intermediate volcanoes, 86 volcanic calderas, 259 arachnoids, and 50 novae. In venusian geological terms, arachnoids are large concentric structures of mysterious origin; scientists derived the name from spider webs, which vaguely resemble their networked appearance. Most appear as concentric circles with a complex, surrounding network of associated fractures. They can stretch as far across as 200 kilometers. The term nova, in venusian geology, refers to radially fractured centers, many of which show evidence of extensive past volcanic activity, and are somewhat raised from the surface, spanning as large as 300 kilometers. They are thought to have formed through upwelling from below and fracturing of the surface, associated with dikes and volcanism.

Some volcanoes on Venus have lobed perimeters that show a splay of outflows of lava from their edges, and this is particularly the case with many of the intermediate examples. Some of them have flower-like, radar-bright flows of lava from their centers, and apparently high central peaks, and others have depressions apparently caused by fault collapses. The areas of Beta Regio, Atla Regio, Themis Regio, and Imdr Regio contain high numbers of them.

Large volcanoes appear more like the Earth analog we are instinctively familiar with, showing large areas of radiating lava flows, a central depression at the summit, and substantial altitude above the surrounding venusian terrain. The margins of the shield that form these big volcanoes are offset nicely from the surrounding landscape, and often these large volcanoes are characterized by a sharp terminus of blocky, radar-bright lava at the end of the flows.

Coronae are among the most abundant structures on Venus; they are found abundantly between the areas of Beta Regio, Phoebe Regio, and Themis Regio, and there is an amazing chain of coronae stretching between Parga and Hecate Chasmata. Five basic types of these coronae exist: concentric, concentric double ring, radial/concentric, multiple, and asymmetric. The largest of them all is a very unusual landform, Artemis Chasma, at 2,600 kilometers in diameter. Coronae are believed to form when a diapir, that is, a molten, ductile blob of hot material, works its way up through the surrounding rock, blurting out finally onto the planet's surface. As it approaches the surface, the hot diapir

Figure 6.6 The volcano Maat Mons stretches five miles high into Venus' thick air in this 3-D image generated from Magellan spacecraft radar data, with simulated colors pulled from the Soviet Venera 13 and 14 landers.
NASA/JPL

causes an upwelling of the surface, and when the blob breaks into the lithosphere, it flattens and spreads out, finally cooling, and forming the disk-shaped corona above it. Then volcanism can take over and sometimes obliterate the shape that was originally formed.

Similar to coronae are features that appear to form in a completely different way and are termed circular depressions. They show concentric faulting, like coronae, but are depressions with floors that can be several hundred meters below the surrounding terrain. They are similar to the inverse of a corona, with floors that often show fracturing. Similarly, calderas are generally circular depressions that contain outcrops of radar-dark, smooth lavas. Several notable examples offer up radial flows of lava away from their edges, as with Sacajawea Caldera, located on Lakshmi Planum. Although these are also depressions in the landscape, the calderas are clearly of volcanic origin and may have formed when magma rose near the surface and then withdrew downward, and the surrounding crust faulted.

Each of these many kinds of volcanic features on Venus offers a window into the planet's tumultuous past. Large and small, featuring lavas that were very fluid or extremely thick, leaving behind smooth plains and rough, blocky landscapes, they cover an enormous range of types, from tiny to many hundreds of kilometers across.

The key common denominator with Venus is that all of these features resulted from mantle upwelling, and planetary scientists know the surface was "redone" substantially some 750 million years ago. Unlike volcanism on Earth, where the great activity occurs in certain zones along lithospheric plates, Venus is a world globally characterized by volcanoes, large and small. But the distribution of major volcanoes is not random: more than 70 percent of them lie on just one half of the planet's surface.

Planetary scientists know that Venus, an extremely warm planet and a nasty place to be, stored enormous amounts of energy deep inside for a very long time after the planet's formation. There is also radioactive decay continually releasing energy. (The surface temperature and internal temperatures have different causes, of course.) The fundamental cause of Venus's strange history is the lack of plate tectonics. Heat was trapped inside the planet and finally had to burst out.

Scientists know that the better part of a billion years ago a huge amount of this banked energy was released, globally resurfacing the planet. No one yet knows what triggered this event, or why it happened exactly when it did. Instabilities deep within the planet conspired – through physical evolution, the laws of physics, and interplay between countless atoms – to let loose and recover our so-called sister planet in a very significant way. Discovering more about why this catastrophic event happened when it did could serve as lessons for our own planet and its future well-being.

Chapter 7
Is Pluto a planet?

Pluto has evoked emotions among astronomy enthusiasts and the general public at large as much as any other astronomical object. This was certainly the case in the early 1930s, following its discovery by farm boy turned astronomer Clyde W. Tombaugh (1906–1997), and its naming after the Greek god of the underworld. The public's embrace of the strange new world seemed to accelerate after Walt Disney created a canine character friend of the cartoon character Mickey Mouse and named the dog Pluto, apparently inspired by the new planet. Certainly, the dramatic story of the discovery of Pluto, by a 23-year-old from Kansas, swelled American pride in the ability of a young kid to make an astonishingly unlikely discovery against the odds through simple persistence and dogged determination.

The emotions over Pluto from the 1930s, however, paled compared with the emotions that followed a controversial decision in 2006. In that year, the International Astronomical Union (IAU), the official organization of astronomers charged with naming and defining cosmic bodies, "demoted" Pluto. For 76 years, Pluto was known to one and all as the ninth planet of the solar system. In 2006, the IAU reclassified Pluto as a dwarf planet, a new categorization, after the realization of some years that many icy bodies exist in the Kuiper Belt, the disk of enormous numbers of small and medium-sized bodies in the region of Neptune and Pluto. Such bodies had been discovered starting in 1977, with the asteroid 2060 Chiron, and many icy asteroids with eccentric orbits were uncovered in the following years, particularly in the 1990s. Astronomers realized they would ultimately find enormous numbers of icy asteroids in the Kuiper Belt, some speculating about tens of thousands, and feared that Pluto was just the most well known of a vast new class of objects.

At a meeting in Prague, a small fraction of the members of the IAU voted on redefining just what constitutes a planet. The IAU produced three major criteria for defining a planet: (1) a planet orbits the Sun; (2) it is massive enough to exist in hydrostatic equilibrium – that is, it is spherical – and (3) it has "cleared the neighborhood" of smaller bodies within its orbit. As defined by the IAU Prague vote, the solar system changed to its present state, containing eight planets –

Figure 7.1 Pluto and three of its moons appear as simple points of light in this Hubble Space Telescope image. Before NASA's New Horizons mission, this was the best image of the system.
NASA

Mercury, Venus, Earth, Mars, Jupiter, Saturn, Uranus, and Neptune; and five dwarf planets, Pluto, Ceres, Haumea, Eris, and Makemake. They even named trans-Neptunian objects large enough to be spherical plutoids. In late 2006, in fact, the IAU "gifted" Pluto with an asteroid designation, 134340 Pluto.

Immediately, in the wake of the IAU decision, an uproar of interest in the planet arose. Schoolchildren seemed miffed at the demotion and wouldn't let the issue go. Amateur astronomers generally were upset, thinking the "demotion" of Pluto ridiculous, as it had been considered a planet for more than three-quarters of a century, and was held in high regard in the hearts of astronomy enthusiasts everywhere as the most distant planet – even if it could only be seen as a faint "star" in backyard telescopes. The world press had a field day with the issue, and it seemingly wouldn't die. To exacerbate the hard feelings of the pro-Pluto set, Caltech astronomer Mike Brown (1965–), leader of the team that discovered 16 Kuiper Belt objects, including Eris, wrote a 2010 book titled *How I Killed Pluto and Why It Had It Coming*. Neil deGrasse Tyson (1958–), director of the Hayden Planetarium in New York City, further stirred the controversy by placing Pluto with other small solar system bodies and away from the planets in the newly remodeled exhibit at the planetarium, just after the IAU vote, fueling dissatisfaction from strangely brokenhearted schoolchildren visiting the institution.

Although my heritage in observational astronomy is strongly tilted toward deep-sky objects – star clusters, nebulae, and galaxies – I will admit to having a soft spot in my heart for Pluto. After all, when I was a young editor at *Astronomy* magazine and editor of *Deep Sky* in the 1980s, I hung around a set of friends and writers that included comet discoverer David H. Levy (1948–), Lowell Observatory astronomer Brian A. Skiff, and others – friends of Clyde Tombaugh (David became his biographer). On a number of occasions, I observed with Clyde, either at his home in Las Cruces, New Mexico, or once for several days at the Texas Star Party in Ft. Davis, when he spoke at the astronomy meeting.

So I liked Clyde a lot and he was great fun. Clyde was then in his 80s, very short and stooped, due to spinal curvature, somewhat frail, but absolutely hilarious – full of rapid-fire puns. He loved talking about various objects in the sky, much more distant than Pluto, and always had funny things to say or do. Once in Texas he watched carefully and assisted as I sketched some deep-sky objects in a large art notebook I had, using a big telescope that necessitated a ladder to reach the eyepiece. (We had to be quite cautious helping Clyde climb ladders in the darkness.) Suddenly, I couldn't find the notebook, and we all started a search for it.

The book had some valuable drawings in it, in the sense that they had taken lots of time to make, and I was a little alarmed. After David Levy and others wandered around with me, looking for it, suddenly Clyde realized he had been sitting on it – it was underneath his bottom and on top of a folding chair. We all laughed for a while, and – given the rather silly nature of the group – Clyde signed the sketchbook, "I sat on this book/Clyde W. Tombaugh/29 May 1987." David tried to one-up him by signing, "It was my chair/David H. Levy/29 May 1987 TSP." I added a final flourish of kookiness by signing, "It's my book/Dave Eicher/29 May 1987."

But I digress. Amazingly enough, the technical status of Pluto became one of the most thought-about aspects of planetary science over the past decade. Pluto has just been visited (in July 2015) by the New Horizons spacecraft, an exciting flyby mission that will reveal enormous amounts of science about the "last frontier" of the major solar system bodies, at least as they were once defined.

What makes the whole Pluto issue still controversial is the nature of the IAU's definition, which was poorly thought through. The problem and the controversy come from the last criterion – a planet must "clear its orbit" of smaller bodies – which is ambiguous at best, if not outright flawed. I have actually heard planetary scientists say, "what does this really mean, even?" According to the IAU, Pluto has not cleared its orbital neighborhood of smaller bodies, and therefore it is a dwarf planet, and not a planet. Clearly, the Pluto argument hinges on this technicality – Pluto does orbit the Sun and is large enough to be spherical. But many planetary scientists remain skeptical of the IAU ruling because they believe it biases against objects of smaller size with increasing distance from the Sun and is squarely at odds with other IAU classification schemes that rely on an object's intrinsic properties, not its location.

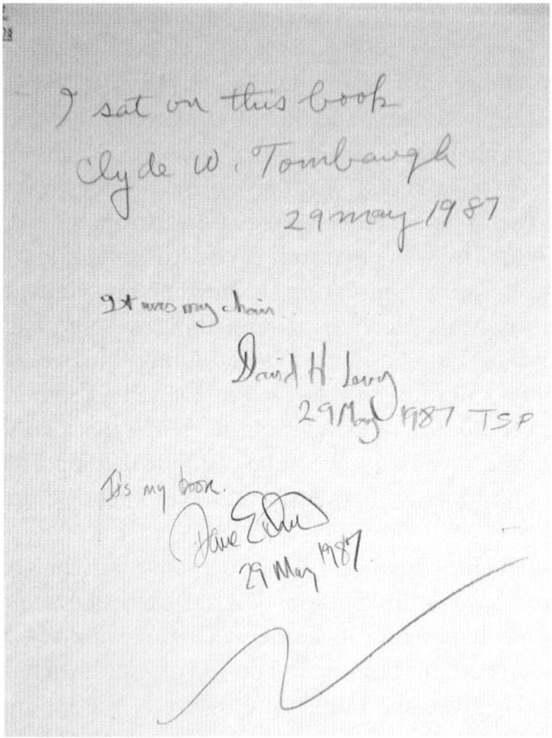

Figure 7.2 A record from David Eicher's sketchbook of Clyde Tombaugh's infectious humor, recorded at the Texas Star Party, Ft. Davis, Texas, May 29, 1987. (See the text for the story behind this artifact.)
David J. Eicher

In early 2014, I had the pleasure of attending a lecture in New York by planetary scientist S. Alan Stern (1957–), principal investigator of the New Horizons mission. He made several salient points about the Pluto controversy. A house is a house whether it stands in a city or in the countryside. Shouldn't a planet be a planet regardless of where it exists? At the Pluto-like distance of 40 astronomical units – 40 times farther away from the Sun than we are now – Earth would not clear its orbit of asteroids, and so would Earth then not be classified as a planet? Planetary scientists have scratched their heads about this, remarking that stars are stars wherever they exist – and, in fact, stars can orbit stars. Some astronomers even openly consider very large moons like Ganymede and the Moon planets as well, and have posed the question, said Stern, "If stars can orbit stars, why can't planets orbit planets?"

To make matters worse, in 2010 astronomers discovered the first Earth Trojan asteroid (confirmed in 2011), 2010 TK$_7$, which orbits at the L4 Lagrangian point in our planet's orbit, leading Earth by 60° in our orbit. It is a tiny asteroid spanning some 300 meters. Further, Earth has a second associated asteroid, 3753 Cruithne, a 5-kilometer rock in a so-called overlapping horseshoe orbit in

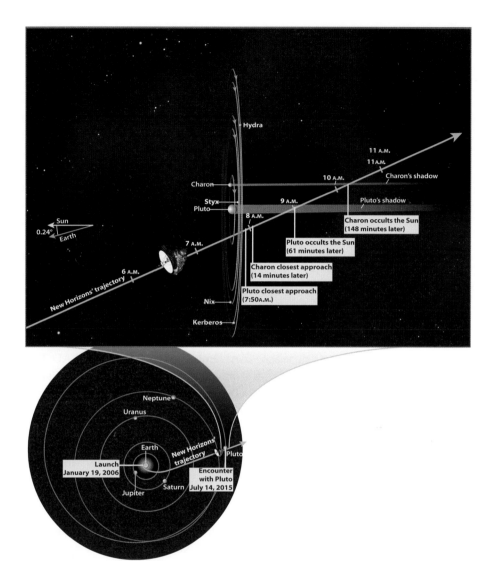

Figure 7.3 Pluto's far-out elliptical orbit takes it nearly 50 astronomical units – 50 times the Sun–Earth distance – deep into the solar system. This graphic shows the New Horizons spacecraft's encounter with the Pluto system that happened in 2015.
Astronomy: Roen Kelly

resonance with Earth. Moreover, five Trojan asteroids of Mars are known, nine asteroids associated with Neptune are known, similar objects are associated with Uranus. Jupiter has a substantial population of Trojan asteroids on both "sides" of the planet in its orbit, as well as the Hilda group of asteroids in orbital resonance.

The IAU definition, then, is so sloppy, vague, and ill conceived that by it one could inquire as to whether Earth, Mars, Jupiter, Uranus, and Neptune are planets, as well as questioning Pluto, as they have not technically cleared their orbits of coexisting asteroids. You can appreciate the mess that this has created. And by the way, Neptune crosses Pluto's orbit, so Neptune is out by that reason alone!

Stern has publicly criticized the redefinition of Pluto. In fact, he suggests that Pluto is indeed a dwarf planet, but that dwarf planets are planets in every measure as well. He suggests that the New Horizons mission, the 2015 flyby encounter and the lengthy, months-long return of substantial amounts of data from the spacecraft, that will last well into 2016, are likely to clarify the issue among scientists and the public at large. Marc W. Buie (1958–) of the Southwest Research Institute has also been openly critical of the classification, as have other influential planetary scientists. The issue seemingly won't go away, and is very much in flux – made worse by further discoveries each year.

Despite the controversy over its planetary status, Pluto remains a fascinating object for science. It is the largest known object in the Kuiper Belt, and the largest known dwarf planet, at approximately 2,368 kilometers in diameter. It is about two-thirds the diameter of the Moon. Dwarf planet Eris is a shade smaller than Pluto, but slightly more massive, due to a greater density. Pluto has the orbit of a classical Kuiper Belt object, at roughly 40 astronomical units from the Sun, approaching as close as 29.7 AU but straying as far as 48.9 AU. By contrast, Eris has a highly eccentric orbit and lies within the scattered disk, a diffuse population of trans-Neptunian objects, farther out than the Kuiper Belt, that have been "spun up" in orbital energy by encounters with Neptune. It strays as far as 97.7 AU from the Sun. Although Pluto is the largest known Kuiper Belt object, it isn't large or massive by our standards. It contains about 0.2 percent the mass of Earth, and stretches about 66 percent the Moon's diameter. Its collective surface area is roughly equivalent to that of South America. Its circumference is equivalent to driving from Manhattan to Moscow – that is pretty sizeable!

Pluto has five satellites, and their discovery stories are amazing. Pluto has a large satellite, Charon, discovered by US Naval Observatory astronomer James W. Christy (1938–) in 1978. Christy examined a high-resolution image of the planet and noticed a "bulge" on one side that subsequent research revealed to be the moon, named after his wife Charlene, and also coincidentally tied to the Greek mythological ferryman who carries souls across the rivers Styx and Acheron. Christy has always pronounced the name with a soft S sound, "sha-ron," rather than a hard K sound, "ka-ron." The moon's existence was confirmed during a series of mutual eclipses and transits between 1985 and 1990.

Charon is very large relative to Pluto, at some 1,208 kilometers in diameter, and so Pluto and Charon form what some astronomers term a double planet, as a moon with half the "planet's" diameter is extremely large. Earth and the Moon are also sometimes referred to as a double planet. Planetary scientists don't know what the origins of the plutonian moons was, but suspect a major impact similar to the huge collision that they believe produced Earth's Moon. In the same way, a significant impact with a large object and Pluto would have created substantial debris that could have accreted into the system's moons.

In 2005, astronomers using the Hubble Space Telescope discovered two additional moons of Pluto, Nix and Hydra. Nix, which may be 45 to 140 kilometers in diameter (depending on its reflectivity), was named after the Greek goddess of darkness and night, the mother of Charon. Hydra, Pluto's outermost moon, perhaps 60 to 165 kilometers in diameter (depending on its reflectivity), was named after the nine-headed serpent who battled Hercules in Greek mythology.

Further, in 2011 astronomers found a fourth moon of Pluto, Kerberos, again using the Hubble Space Telescope. Kerberos has an estimated diameter of 13 to 34 kilometers (depending on its reflectivity) and was named after the Greek form of the multiheaded mythological dog who guards the entrance to the underworld. A year later, the Hubble team discovered a fifth plutonian moon, Styx, perhaps 10 to 26 kilometers across (depending on its reflectivity) and named after the Greek goddess of the river Styx in the mythological underworld.

Despite the intense public furor over Pluto of the distant past – and the recent past – we really know very little about Pluto as a physical object. Thus, the hot interest in the New Horizons mission. Astronomers have studied Pluto using Earth- and space-based instruments since the 1950s, with a variety of techniques – photometry, spectroscopy, and polarimetry. They initially focused on very basic properties like Pluto's color and its albedo, its surface reflectivity. Almost right away, astronomers found that Pluto's albedo is higher than Charon's. Recent observations with the Hubble Space Telescope have revealed that Charon varies in brightness by about 8 percent as it rotates, but Pluto also varies by 25 percent in reflectivity as it rotates. Charon has a relatively low range of contrast, but Pluto has darker and lighter areas arrayed across its surface and strong contrast.

Pluto's color has been a subject of study since the 1950s. Astronomers have found it to be reddish, slightly more so than Neptune's moon Triton, but not as red as the refractory surfaces of Mars, which are stained by copious amounts of iron oxide. Planetary scientists believe Pluto's color results from the creation of complex organic molecules by radiation acting on the nitrogen–methane–carbon monoxide surface ices.

Deciphering the surface composition of Pluto has been a longstanding challenge to planetary scientists. Since the 1950s, they have made observations in the infrared part of the spectrum, but Pluto's incredibly feeble light as seen from Earth meant that little progress could be made until the 1970s. In 1976, astronomers discovered methane ice absorptions in Pluto's spectrum. In 1992, the spectroscopic study of Pluto took another leap forward when astronomers used the UK Infrared Telescope on Mauna Kea, Hawaii, to detect nitrogen and carbon monoxide ices on Pluto. The year 2006 brought the discovery of ethane ices on Pluto's surface, and this type of hydrocarbon had long been suspected because of the processing of Pluto's surface and atmosphere by chemistry and

Figure 7.4 The New Horizons spacecraft races past Pluto in this artist's illustration.
Johns Hopkins University Applied Physics Laboratory/Southwest Research Institute

radiation; more evidence for hydrocarbons was obtained using the Hubble Space Telescope in 2012. Astronomers also believe that Pluto's darker regions could contain more highly processed hydrocarbons.

From their generation of research, planetary scientists believe Pluto is a frozen world with widespread methane, found in higher concentrations where the surface is more reflective. (The methane is widespread but not abundant; the surface contains 10 times more nitrogen than methane.) They believe the methane is dissolved with other ices, but sometimes observed as essentially pure methane ice. Abundant nitrogen and carbon monoxide also exist. Ethane, evolved from chemical reactions that began with methane, also exists. Rocky material could also be present, and certainly lots of processed hydrocarbons exist due to long-term radiation exposure. A surface frost consisting of volatile ices also exists.

Just how cold are those ices? Measuring Pluto's surface temperature is also a tricky challenge. The first good attempts came from the Infrared Astronomical Satellite (IRAS) in 1983. IRAS measured Pluto's temperature at about 60 K, or −213 °C (−351 °F). But planetary scientists soon found a more complex situation with regard to Pluto's surface, one that required some retooling of the data. Researchers found that a large portion of the plutonian surface is actually far colder than this, in the vicinity of 35 K to 40 K, or about –233 °C (–387 °F). Subsequent studies revealed the fact the Pluto must have surface temperature variations such that some areas are warmer than others. Studies with instruments, including the Spitzer Space Telescope confirmed this and suggest that colder areas of the planet are those that are more reflective, richer in ices, and warmer areas are darker, devoid of larger amounts of highly reflective ices.

Studying features on Pluto's surface has been a monstrous challenge for astronomers, as the planet spans less than 0.1 arcseconds on our sky, and this will no doubt be one of the most exciting aspects of the New Horizons mission. Nonetheless, astronomers have created relatively crude maps of the surface brightness of Pluto with the Hubble Space Telescope. Indeed, until Hubble was available, Pluto's disk could not even be resolved from Earth. Despite this, in the 1950s evidence for surface markings on Pluto existed because of the strong rotational light curve, and later astronomers produced good maps from light curves of mutual events, occultations, and transits of Pluto and Charon.

More recently, astronomers have studied Pluto's appearance with instruments that are more sophisticated. Two significant teams have studied Pluto's light curve in recent years, one led by Marc Buie at Lowell Observatory (using the Hubble Space Telescope) and the other led by Eliot Young of the Southwest Research Institute and Richard Binzel of the Massachusetts Institute of Technology (studying mutual events). These groups used different techniques that enabled their research to be cross-checked to form a composite result. They found commonalities that included a bright south polar cap, a dark band over mid-southern latitudes, a bright band over mid-northern latitudes, a dark band at high northern latitudes, and a bright northern polar region.

In 1994, astronomers first used Hubble to image Pluto and create a composite surface map. They delivered more data showing that Pluto's surface is highly variegated in brightness; has extensive, bright, asymmetric polar regions; has large spotty features at mid-latitudes and along the equator; and shows what appear to be large linear features stretching hundreds of kilometers long. Amazingly, the range of brightness in albedo features on Pluto's surface is about 5 to 1, meaning that some areas are 5 times brighter than others – Pluto has more contrast on its surface than any objects in the solar system other than Iapetus and Earth. Further Hubble imagery taken in 2002 and afterward built on these results.

Deciphering the internal composition of Pluto is a trickier matter altogether. The most basic measurement is that of Pluto's density, which requires very precisely measuring Pluto's position – its wobble as it and Charon and the other moons orbit a common center of mass. Astronomers have used the Hubble Space Telescope to determine this, but it's a challenge – the best result came from analyzing the motions of Nix and Hydra, and yields a density for Pluto of about 2 grams per cubic centimeter, less than half of Earth's density.

Low as it is relative to Earth, astronomers expected Pluto to have an even lower density, something approaching water's reference of 1 gram per cubic centimeter. The most confident modeling of Pluto's composition suggests it contains primarily three constituents: water ice, rock, and methane ice. Pluto's density of 2 grams per cubic centimeter means that Pluto contains a high volume of rock, something of the order of 70 percent, similar to the rocky content of the moons Io, Europa, and Triton. The relatively high fraction of rock believed to be in Pluto suggests the nebular material that formed Pluto was quite rich in carbon monoxide rather than methane.

The nature and layering of Pluto's insides are unknown, and a topic of great interest to planetary scientists. Practically all of the factors influencing how material is organized within Pluto are unknown: viscosities, fractions of rocky materials, the internal content of radioisotopes, the distribution of various densities of material – it really is a blank slate. But knowing what they do know, planetary scientists suggest that perhaps Pluto exists in one of two basic models. The first postulates a core of partially hydrated rock, surrounded by a mantle of ices 400 kilometers below the surface, and a crust of water ice forming a shell some 135 kilometers thick. The second model proposes a core of partially hydrated rock within a mantle of organic materials around 350 kilometers below the surface, and a crust of water ice 250 kilometers thick. But they are hamstrung by not knowing how materials are differentiated within Pluto. They do believe that Pluto's interior reaches temperatures of 100 to 200 K (–173 °C or –279 °F to –73 °C or –100 °F), which may indicate the presence of an internal ocean.

Pluto's thin atmosphere is another compelling aspect of the planet, and one that will be a fascinating target of the New Horizons spacecraft. In 1976, astronomers discovered methane on Pluto, and that hinted strongly toward an atmosphere because of suspected vapor pressures at the temperature on Pluto's surface. The high albedo of Pluto's surface and "spots" of lightness also suggest some resurfacing on the object, which also indicate potential atmospheric activity. But it took until the late 1980s to find clear evidence of a plutonian atmosphere.

In 1988, Pluto occulted – passed in front of – a 12th-magnitude star, which allowed astronomers to make very precise measurements of the effects of the plutonian atmosphere on the star's light. American astronomers Robert Millis of Lowell Observatory and James Elliot (1943–2011) of the Massachusetts Institute of Technology (MIT) led a team that used the Kuiper Airborne Observatory to observe the occultation. (Elliot had previously discovered the rings of Uranus with an observing team in 1977.) When Pluto passed in front of the star, the star's light faded more gradually than it would have had Pluto been without atmosphere – the distinct signature of atmospheric absorption.

Subsequent theoretical studies by various astronomers produced models for the temperature of Pluto's atmosphere, suggesting a value of 106 K (–167 °C or –269 °F). This relatively high temperature compared with the planet's surface is because the efficiency of radiation and cooling is small. Scientists know that nitrogen dominates Pluto's atmosphere, and that carbon monoxide and methane are trace gases.

In 1994, MIT astronomers directly detected methane in Pluto's atmosphere. The detection came through high-resolution infrared spectroscopy and points to a very small amount, an abundance of possibly 10 to 1 percent of the total.

Knowing which gases probably make up Pluto's atmosphere is one thing; modeling how the atmosphere is put together – its structure, altitude above the surface, and differentiation – is a much more difficult proposition. The 1988 occultation suggested that at an altitude of 1,215 kilometers the refractive index of the atmosphere changed sharply. One explanation for this is that aerosol hazes may exist in Pluto's atmosphere. These could be photochemical "smogs" similar to hazes that astronomers have observed on the moons Titan and Triton. Or, the change in atmosphere at this altitude could be caused by a sudden change in atmospheric temperatures.

More occultation events took place in 2002, and astronomers studied the plutonian atmosphere carefully once again. They did not find this differentiation at 1,215 kilometers in 2002, suggesting that major changes took place in Pluto's atmosphere in the interval between the observations, a span of only 14 years. Observations of still more occultations in 2006 suggest the planet's lower atmosphere had become more turbulent in the recent several years. New Horizons data will no doubt provide a much clearer picture of what's happening with Pluto's atmosphere.

Pluto's moons are of special interest to astronomers because of the strange dynamic of the system, with so many small moons orbiting Pluto, and, of course, because of the nature of Charon's enormous size relative to Pluto. Charon's diameter of about 1,208 kilometers makes it about half that of Pluto, or a little bigger than half. Charon's mass is in the range of 10 to 14 percent that of Pluto's, whereas typical mass ratios of planets to moons are about 1,000 to 1. Charon's large relative size is a clue about how the system formed.

Mutual events between Pluto and Charon, observed between 1985 and 1990, uncovered some basic facts about Charon. Its albedo is about 30 to 35 percent, making it much darker than Pluto, though still highly reflective. It also has a much more neutral color as opposed to Pluto's quite reddish hue. Spectral analyses of Charon's surface ices show it does not have the rich signature of methane that Pluto shows. It does show evidence of water ice, and it is possible that Charon lost its coating of volatiles from impact-related heating or from the loss of a thin atmosphere. Alternatively, Charon may have formed without the same volatiles as Pluto.

At perihelion, Charon has a surface temperature of about 60 K (–213 °C or – 350 °F), meaning it is probably too cold to have an atmosphere, given its water ice dominant surface composition. In 2001, astronomers using infrared telescopes on the ground detected ammonia or ammonium hydrates along with crystalline water ice on the surface of Charon. These results suggest that some type of geological activity may be present on Charon.

The other moons are far less known, and certainly New Horizons will flood us with a trove of data. Nix orbits Pluto over a period of about 25 days at a distance of roughly 49,000 kilometers. Little is known about it beyond its diameter, which ranges from some 45 to 140 kilometers. The moon appears to have a surface that is essentially gray in color. Hydra orbits Pluto once every 38 days at a distance of about 65,000 kilometers, and is similar in terms of its spectrum to Nix, suggesting a grayish color, with a size range of 60 to 165 kilometers. Pluto's fourth moon, Kerberos, has a 32-day period and orbits at a distance of approximately 59,000 kilometers, and may be between 13 and 34 kilometers in diameter. The fifth moon, Styx, has a period of 20 days at an orbital distance of 42,000 kilometers, and it is probably 10 to 25 kilometers in diameter. Will the New Horizons flyby uncover other, tiny Plutonian moons?

Planetary scientists have made strides recently in understanding the possible origins of the system of satellites orbiting Pluto. The bodies, chiefly Pluto and Charon, could have accreted together in the solar nebula. They might have been formed following an impact between the proto-Pluto and proto-Charon that cast the material for the satellites into orbit around Pluto. Charon may have formed by rotational fission; that is, rotationally cast out of Pluto. The first and last possibilities are remote – accretion isn't very likely given the bodies' small sizes, and rotational fission is unlikely on grounds of unrealistic angular momentum.

That leaves the most likely idea about the formation of Pluto and Charon as an impact hypothesis, wherein Pluto and proto-Charon were on intersecting orbits and slammed into each other. This is an analog to the Earth–Theia–Moon creation hypothesis that likely explains the origin of the Moon. Both proto-Pluto and proto-Charon most likely accreted from small planetesimals and then met in a big collision that resulted in the system we see today.

In 2005, Robin Canup, whose work on the lunar Giant Impact Hypothesis was significant, published her results on a study of the Pluto system. Her work suggests that Pluto likely collided in an oblique fashion with another body that had 30 to 100 percent the mass of Pluto, and that the collision probably took place at a velocity of about 1 kilometer per second. And nearly 20 years beforehand, William B. McKinnon of Washington University in St. Louis, Missouri, produced pioneering work on this idea. Also in 2005, the discovery of the tiny moons Hydra and Nix added momentum for the Giant Impact Hypothesis. The orbital plane being the same for all the moons is taken as supporting evidence of the collision, as is a nearly perfect resonance on orbital periods.

The mounting evidence for Pluto's satellite system forming in a giant impact scenario adds to the understanding planetary scientists now have of the enormous number of icy bodies in the Kuiper Belt. The icy nature of Pluto and the type of constituent compounds in its spectrum strongly suggest that it – and other similar, icy bodies – formed in the outer solar system. But Pluto's small size and the vast amount of space in the outer solar system, coupled with the almost certain collision scenario, leaves an element of mystery. How could such small bodies slam into each other with so much open space in the region?

Figure 7.5 This illustration shows Sedna, a near Pluto-sized object in an orbit that takes it deep into the reaches of the outer solar system. A family of dwarf planets discovered at the solar system's edge has changed the way astronomers view our planetary neighborhood.
NASA/JPL-Caltech/R. Hurt (SSC)

In 1991, Alan Stern proposed the existence of a large group of small, icy bodies in the region of Pluto – some hundreds or thousands of ice dwarf planets that formed during the accretion of Uranus and Neptune, at a distance of 20 to 30 astronomical units from the Sun. This population could help explain the collision that astronomers believe formed the Pluto–Charon system, as well as the capture of Neptunian moon Triton, and the strange axial tilts of Uranus and Neptune.

The large number of bodies discovered in the Kuiper Belt since 1992, now numbering more than 1,000, is still only the tip of the iceberg. Astronomers believe that at least 100,000 Kuiper Belt objects with a diameter of more than 100 kilometers must exist in the region between 30 and 50 astronomical units from the Sun. Pluto, Haumea, and Makemake are simply the largest known members yet discovered, and Eris lies a little farther out in the so-called scattered disk, where such small icy objects have had their energies "spun up" by encounters with Neptune, and have crazier orbits with greater inclinations and eccentricities.

Astronomers also believe the Kuiper Belt and the scattered disk were far more heavily populated in the early days of the solar system. More than 80 years after Pluto's discovery, scientists now see this object, its large moon Charon, and the other small moons in proper context. It did not form in isolation, but it is

Figure 7.6 NASA's New Horizons spacecraft captured this view of Pluto during its flyby 14 July 2015.

rather a relic of an earlier age, and it still remains whereas most of the actors in that early drama have left the stage, cast out into the far reaches of the solar system or beyond. A vast population of small planets once inhabited the Kuiper Belt. Pluto remains today in large part because of its 2:3 orbital resonance with Neptune, which other so-called Plutinos also have.

We now see Pluto clearly, for the first time in context. It is a dwarf planet – *and* a planet. It is one of the most crucial small objects in the solar system, opening a gateway on understanding icy planets and on a huge era of discovery that will no doubt come. And we are perched on the doorstep of an explosive dawn of understanding Pluto and all small bodies in the solar system as New Horizons swings past the planet in 2015 and delivers its data for a span of more than 16 months. Wherever he is, despite the International Astronomical Union, Clyde Tombaugh is smiling.

Chapter 8
Planets everywhere . . .

Few areas of astronomy or astrophysics have been as explosive over recent years as the cottage industry of extrasolar planet discoveries. At the time of writing in early 2015, astronomers know of more than 1,800 planets orbiting more than 1,100 stars other than the Sun, all relatively nearby in the Milky Way Galaxy. The most productive discoverer of exoplanets, NASA's Kepler spacecraft, has produced a total of some 1,000 confirmed exoplanets and 2,900 exoplanet candidates that await confirmation with additional data. The pace of discoveries and confirmations has been so dizzying – especially during the busiest periods of analyzing data from various instruments dedicated to the task – that the count of exoplanets has changed from week to week.

Speculations and crude research on the existence of planets beyond the solar system go back a long way. In the sixteenth century, Italian philosopher Giordano Bruno (1548–1600) speculated on the "infinity of worlds" that should exist out in the vastness of the stars. Similarly, in the conclusion of his fantastic tome *Principia Mathematica*, Isaac Newton suggested that stars are the centers of systems like the Sun and planets.

The first claims of a detection of exoplanets rolled along in the nineteenth century. Claims of detecting a "dark body" affecting the orbit of the star 70 Ophiuchi, a 4th-magnitude star lying 16.6 light-years distant, stretch back to 1855, when observers at the East India Company's Madras Observatory reported wobbles in the star's position in the sky. Similarly, American astronomer Thomas Jefferson Jackson See (1866–1962), an eccentric underachiever, claimed observations that proved such a body in 70 Oph throughout the 1890s. But these were subsequently discredited. Sixty years later, Dutch astronomer Peter van de Kamp (1901–1995), working at Swarthmore College's Sproul Observatory, claimed detection of a planet orbiting another close star, Barnard's Star. This star, also in Ophiuchus, has the highest degree of motion relative to the background stars in the sky, at a distance of only 6.0 light-years. But the van de Kamp observations also proved to be erroneous.

The first confirmed detection of a planet outside the solar system was announced in 1992, when Polish–American astronomer Aleksander Wolszczan

(1946–) and Canadian–American astronomer Dale Frail described their detection of two planets orbiting the pulsar PSR B1257+12. They found a third planet in the system 2 years later, and, amazingly, the astronomers believe the planets formed in the wake of the supernova explosion that formed the pulsar, or perhaps the planets were all that remained of the cores of gaseous planets that had been stripped of their lighter elements in the supernova event.

The first detection of a planet orbiting a main-sequence star, which resonated with the press in a larger way, came along in 1995. Late in that year, Swiss astronomers Michel Mayor (1942–) and Didier Queloz (1966–) announced their discovery of an extrasolar planet orbiting the star 51 Pegasi, a sunlike, G-type star shining at magnitude 5.5 and lying some 51 light-years away. The astronomers used the 1.93-m telescope at the Observatorie de Haute-Provence in France, along with the ELODIE echelle spectrograph, to detect a wobble in the star's radial velocity, betraying the existence of a planet. Within a short time, the planet-hunting team led by American astronomers Geoff Marcy (1954–) and Paul Butler in California confirmed the discovery. Subsequent research led to the belief that the planet orbits very close to 51 Peg, has a mass about half that of Jupiter, and suffers from extraordinarily high temperatures. It was the first of a series of similar planets to be found later, which came to be known as "hot Jupiters."

In the wake of the explosive 51 Pegasi announcement, which garnered worldwide media attention, astronomers began to search for more exoplanets, with an increasingly sensitive set of instrumentation and with a variety of techniques. They also carefully pondered the question of where, exactly, they should be looking.

What kinds of stars could host habitable planets, and perhaps even life? Astronomers believe that O and B stars are the only types that can't host life-bearing planets, because of their relatively short lifetimes. G stars like the Sun are an obvious choice to analyze, as our only example of life in the cosmos exists on a planet, ours, orbiting a G star. Only about 7 percent of the stars in the Milky Way are G stars, but that still amounts to billions of stars. Small and dim stars in the K and M spectral classes might have very small habitable zones – the temperature regions where liquid water could exist – but it is unlikely that any planets they may have could host life. But astronomers don't yet really know. The infrared radiation emitted in quantity from M dwarfs would be efficient in terms of warming a planet by the greenhouse effect. It is hard to say how this will play out in the end. These small, dim stars are the most numerous types that exist, and so even if life is rare among them, they still offer many places to search for it.

The orbital complexities of multiple star systems might seem at first blush to rule out the possibility of hosting planets. The binary star's separation is key here, as it is with a single star. The fraction of stars with planets would be low for binary stars separated by 10 AU, moderately lower for binaries separated by 10 to 100 AU, and almost no different from single stars for binaries separated by more than 100 AU. Some planets could orbit both stars by being far away, too.

More than half the stars in the Milky Way are in binary or multiple star systems, and 60 percent of sunlike stars reside in star systems with more than one sun. Binary star systems offer three ways in which a planet could orbit in a stable configuration. First, a planet could orbit both stars at a relatively large distance. Second, the planet could orbit one star with the orbital radius being much smaller than the separation between the two stars. Third, a planet could orbit permanently between the two stars. But in this last case, the orbital distance would have to be just right to maintain any possibility of a stable orbit, or the planet would experience gravitational tugs from each star that would eventually pull it out of a stable orbit. Most likely, this last possibility would be highly unstable and would not last. The stability of planetary orbits within multiple star systems, with three or more suns, becomes progressively trickier. It becomes quickly difficult to imagine planets remaining over very long periods in many multiple star systems, although certainly it is possible in a percentage of them.

The rapidly increasing sophistication of exoplanet searches led to a variety of techniques for detection, all of which have played a part in discoveries. Since 1995, the hunt for exoplanets has transformed into an international effort involving teams of researchers at various institutions. Regardless of the methods used, they are either searching for extrasolar planets directly, as with making images or taking spectra of the planets themselves, or indirectly, as in the detection of a star's properties that give away the presence of a planetary companion, or more than one planetary companion, without detecting the planets themselves. A method of finding planets directly would always be helpful, but telescopes and detectors are just beginning to reach the level of sensitivity that makes this possible. (And spectra of exoplanet atmospheres would be really great, but that's a long way off.) Most planets that have been found have been detected indirectly.

The most powerful method thus far has been using gravity to betray the presence of a planet orbiting a star. When the planet orbits, its gravity tugs a little bit on the parent star, creating a wobble in the star's radial velocity. Most planets detected thus far have been found using this technique. It is possible because the planet is not really orbiting the star; in fact, both star and planet or planets are orbiting a common center of mass that is slightly off from the star's center. In our own solar system, for example, Jupiter causes the Sun to wobble slightly, orbiting the pair's common center of mass once every 12 years. Seen from afar, extraterrestrial astronomers could deduce our solar system has a Jupiter-mass planet at its distance from the Sun by observing the Sun's wobble around this center of mass. Of course, the other planets affect the Sun's wobble too, although to a much lesser extent than Jupiter does.

Astronomers have employed two different techniques to find the wobbles left by gravity's trail – astrometry and the Doppler technique. Astronomers have made astrometric measurements of a star's position, intended to be of very high precision, for decades. It was the technique used in some of the earliest attempts to detect planets orbiting nearby stars, and did not work well. The technique works really well with binary stars, as generally even a low-mass star has a pretty good wobble effect on its companion. But with planets, the effect is more subtle. At a distance of 10 light-years, a Jupiter-mass planet would cause a wobble of 0.003 arcseconds, just within the current limits of detection.

But astrometry is tough because of two factors: first, the farther away a planet is from its star, the smaller the wobble will be; second, the time needed to detect a star's motion can be great, making detection really difficult. It turns out astrometry works best for massive planets that are a fair distance from their star. And because of the laws of orbital motion, this means a planet that takes quite a while to orbit the star, and therefore to show the wobble. Astrometry has not yet unquestionably detected any of the more than 1,800 exoplanets found thus far.

More successful as techniques go is the Doppler, or radial velocity, method. The Doppler technique offers a way to detect the wobble of a parent star by looking not for side-to-side motion of a star as it moves across the sky, but for carefully analyzing a star's spectrum for signs that the star is orbiting a center of mass that includes planets. The Doppler effect is easy to experience with sounds on Earth, as when an ambulance passes you and the sound suddenly switches pitch as it moves from coming toward you to going away from you.

The same thing happens with light. If an object in the cosmos is moving toward us, its light is shifted toward shorter wavelengths, toward the blue end of the spectrum. If an object like a star is moving away from us, its light is correspondingly shifted toward the red. This so-called blueshift or redshift is easy to see in the spectrum of a star, and so this becomes a planetary detection technique. If an orbiting planet causes its host star to move slightly toward us and away from us in an alternating, periodic fashion, that dance-wobble will show in its spectrum. The first exoplanet, 51 Pegasi, was discovered in this way when its motion showed a 4-day period and an orbital speed of some accompanying body of 57 meters per second.

The Doppler technique measures the orbital velocity of stars to within a meter per second, about as fast as you could walk through the park. The technique also allows astronomers to deduce some properties of a planet's orbit. They can also infer something about the masses of planets accompanying a star. So the picture is somewhat incomplete. The Doppler technique only works when a planet is orbiting from some orientation other than face-on to our line of sight. And the Doppler shift reveals the star's orbital velocity only if it is viewed edge-on. So the precision of what astronomers learn from Doppler shift data is not perfect.

The Doppler technique is at its best when used to find massive planets that orbit relatively close to the parent star. The Doppler technique is powerful, and has resulted in the discovery of many exoplanets thus far.

Another method of finding exoplanets that offers great success to astronomers is the transit method. This eliminates concentrating on the gravitational tugs of planets on their stars and instead views stars for considerable periods so

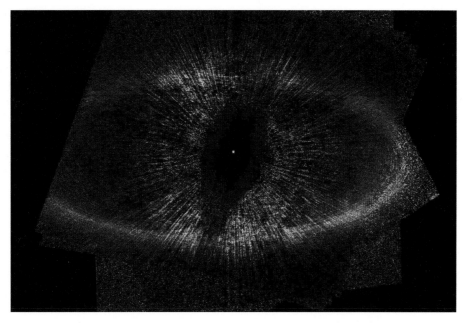

Figure 8.1 A disk of dusty debris surrounds the star Fomalhaut in this Hubble Space Telescope image.

NASA, ESA, P. Kalas and J. Graham (University of California, Berkeley) and M. Clampin (NASA/ GSFC)

the planets can be seen moving across the star's face, dimming their light for a short time. This technique depends on the orbital radius of the planet, and about 1 in 1,000 will have an alignment relative to us such that their planets will cross the star's disk. The small black dot that crosses the star, as with our own solar system's transits of Venus and Mercury, can be captured and analyzed. We cannot photograph the effect of a planet against its star, as the star's disk is far too small at the enormous distances, but the dip in light allows the planetary characteristics to be calculated, given repeated observations.

Discovering exoplanets by transit requires many observations, because stars can dip in brightness for reasons other than a transiting planet. To be sure the star's dimming is resulting from a planet passing in front of the star, astronomers need to see the same dimming multiple times, with the right regularity. A regular periodicity like this then can reveal the orbital period of the planet, and from that astronomers can deduce the orbital distance and mass of the planet. The percentage of light lost in the transit can also point toward the size of the planet. This is tricky, too. An Earth-sized world would create a drop in the star's brightness of just one-one hundredth of 1 percent, which is difficult to measure unless the telescope is orbiting outside Earth's atmosphere.

These techniques all work toward discovering exoplanets that orbit close to their host stars. But there are other methods, too. Gravitational lensing, the same technique that makes very distant galaxies visible when they are lensed through the mass of closer objects, enables spotting exoplanets. Advantages here would be the ability to find small planets like Earth even from great distances through the galaxy. Astronomers focus on candidate stars, monitoring them over long periods, and try to catch one star passing in front of another in the crowded regions toward the Milky Way's center. The stars' alignment is temporary, a result of random motions of stars near the galaxy's center, and when they align, the nearer star acts as a lens, increasing the brightness of the more distant star. Any accompanying planet can also act as a lens, further brightening the star. Several planets have already been discovered using this ingenious approach.

Astronomers are also anxious to detect extrasolar planets directly. But this trick is a towering one, both because of the enormous distances to other stars, and because of the incredible ratio of brightness of the host stars compared to their very dim planets. A sunlike star, in fact, is a billion times brighter than any of its attendant planets. A few planets have been imaged directly, despite these challenges. They are Jupiter-like planets that lie far away from their stars, thus enabling the imaging.

The last few years have witnessed a revolution in exoplanet research. Quite a lot of this has happened with spacecraft missions. Canada's first space telescope, MOST (Microvariability and Oscillations of STars), launched in 2003, observed 55 Cancri e, which had been discovered on the ground, in transit. In 2006, astronomers using the Hubble Space Telescope commenced a study called SWEEPS (Sagittarius Window Eclipsing Extrasolar Planet Search), discovering 16 exoplanet candidates. From 2006 through 2013, an international mission led by the French Space Agency, CoRoT (COnvection ROtation et Transits Planétaires), became the first spacecraft dedicated to discovering exoplanets through the transit method. Its first detection came in 2007 and over the following 6 years it studied more than two dozen planets, and uncovered an additional 600 candidate planets.

In 2008, the EPOXI spacecraft, en route to an encounter with Comet 103P/ Hartley, was used for studying previously discovered extrasolar planets. In 2013, the European Space Agency launched the Gaia spacecraft, designed for high-resolution astrometric mapping of as many as a billion stars in the Milky Way. Additionally, Gaia has the capability of detecting numerous exoplanets, and mission planners hope it will measure the orbits and inclinations of thousands of such planets, thereby determining their masses very accurately.

But the king of exoplanetary discovery has been another spacecraft, NASA's Kepler mission, which was launched in 2009 with the specific purpose of discovering planets orbiting other stars – in fact, to discover earthlike planets if possible, and to derive an estimate of how many such planets within habitable zones might exist in the Milky Way. The results from Kepler have been nothing short of incredible. Kepler utilizes a photometer that continually monitors the brightness of more than 150,000 stars within a $10°$-square patch of sky on the border between the constellations Cygnus and Lyra, and its reach is relatively short – in other words, it can detect planets at most a few thousand light-years away, a small distance relative to the 100,000+–light-year diameter of the Milky Way's disk.

Despite these limitations in sky coverage and distance, Kepler and the associated follow-up observations using earthbound telescopes found 1,000 confirmed extrasolar planets in more than 400 stellar systems, and 2,900 additional planetary candidates. The spacecraft operated normally between 2009 and 2013, when a second reaction wheel failed, and the mission's ability to aim accurately was compromised. A low-cost mission, Kepler had been designed to last 3.5 years. In late 2013, Kepler scientists announced their creation of a new phase of the mission, named K2, that commenced in 2014 and allows the now-disabled spacecraft to carry on, letting the sky rotate into the spacecraft, and using the instruments to detect exoplanets around red dwarf systems.

Kepler's incredible success with detecting exoplanets was a signal triumph. Consider its numbers: about 1,000 planets found and more than another 3,000 suspected in a small area of sky and only looking outward some several thousand light-years maximum. Now extrapolate those numbers if taken all over the sky, and looking throughout the entire Milky Way Galaxy – the numbers would be enormous! In late 2013, the Kepler team announced they could extrapolate, from their data, an estimated figure of as many as 40 billion earthlike planets orbiting within the habitable zones of sunlike and red dwarf stars within the Milky Way. The figure is staggering, and begins to give us an approximation of the numbers of habitable planets that could exist, more or less

Figure 8.2 Kepler-186f resembles Earth perhaps more than any world yet discovered, but orbits a red dwarf star.
NASA Ames/SETI Institute/JPL-Caltech

Figure 8.3 The gaseous planet Kepler-35b is suspected to be in its stars' habitable zone. The world circles a "twin-sun" binary system in a 131-day orbit estimated to be about 60 percent of the distance between Earth and the Sun.
Lynette Cook

like our own world. And remember that is just within the Milky Way – that could be multiplied by 100 billion to get a rough estimate of the numbers of habitable planets in the entire observable universe. It is both a sobering and exciting prospect!

Multiple research teams are working on examining Kepler's planetary candidates with instruments on the ground. And several successive space missions focused on exoplanets are planned for the future. They include CHEOPS (CHaracterizing ExOPlanets Satellite), a European mission slated for a 2017 launch. This space telescope will consist of a 33 centimeter telescope in a 3.5-year mission designed to search for shallow transits, revealing super-Earths (earthlike planets larger than our own) to Neptune-mass planets. Also planned for a 2017 launch is NASA's TESS (Transiting Exoplanet Survey Satellite), an instrument that will employ an array of wide-field cameras to record an all-sky survey of nearby stars for exoplanets.

Further, the James Webb Space Telescope (JWST), the next very large space observatory, will include exoplanet research as part of its wide-ranging science goals. Slated for launch in 2018, JWST's 6.5-m mirror will target exoplanets in a wide range of programs designed to discover exoplanets and analyze known systems. A planned European mission, PLATO (Planetary Transits and Oscillations of stars), should launch in 2024 and search for rocky exoplanets around red dwarfs in the galaxy.

With such an explosion of exoplanet discoveries over the past decade, astronomers have a good general inventory of nearby planets, keeping in mind the bias that it is much easier to find large planets as opposed to small ones. So as instruments and techniques improve, astronomers will certainly find far more and smaller planets, and presumably ones that closely resemble Earth.

For the moment, the gallery of known planets is already pretty spectacular. Some of the most unusual have the most alluring tales to tell. In 2013, astronomers found three exoplanets orbiting the star Gliese 667C in Scorpius, a system lying some 22 light-years from Earth. This remarkable star holds three planets that orbit within its habitable zone, and all three are so-called super-Earths, with masses larger than Earth but no larger than about 15 times greater, which would be about the mass of Uranus. These three habitable worlds are cousins to three and possibly four other planets in this solar system, and the star, Gliese 667C, has about one-third the Sun's mass and only half the surface temperature of our star. So the three planets in the star's habitable zone orbit close in, at a distance that would place them inside Mercury's orbit in our solar system.

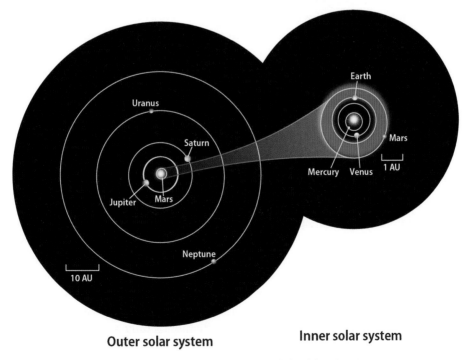

Outer solar system   **Inner solar system**

Figure 8.4 This illustration shows the range of Earth's habitable zone. Astronomers are finding more and more exoplanets, but only a small number of those lie at distances from their host stars where liquid water might exist on their surfaces.
Astronomy: Roen Kelly

Another star with multiple known exoplanets is Kepler-11, a sunlike star in Cygnus lying at a distance of 2,000 light-years. Kepler-11 has a group of planets that are packed in tightly, six worlds orbiting the star within the orbit of Venus in our solar system, and all of which were discovered in 2011. The five closest planets in the Kepler-11 system would fit inside the orbit of Mercury. These planets are all super-Earths, ranging from 1.8 Earth diameters to 4 times our planet's size, and 8 times the mass. The host sun, Kepler-11, is very similar to the Sun, with 96 percent of its mass, very closely the same luminosity, and about the same diameter. It is older than the Sun, at 8.5 billion years, and appears as a faint glow in our sky, at magnitude 14, near the limit of visibility in medium-sized amateur telescopes.

By contrast, another Kepler system, Kepler-64b, is an oddball. Discovered in 2012, it is the first known exoplanet orbiting a quadruple star system, one that is located in Cygnus at a distance of 5,000 light-years. This is a giant planet, Neptune-sized, with a mass in the range of 20 to 55 Earth masses. The planet orbits an eclipsing binary star, consisting of an F-type star with 1.5 times the Sun's mass and an M dwarf with less than half the Sun's mass. The planet orbits this pair of stars once every 138 days. Another binary star is gravitationally linked to this system, and consists of a sunlike star paired with another M dwarf. The two pairs of stars are separated by 1,000 astronomical units – a thousand times the Earth–Sun distance.

Some of the exoplanets discovered thus far have extremely strange properties. For example, 55 Cancri e is an exoplanet discovered in 2004, orbiting a 6th-magnitude star in Cancer, a system lying at a distance of 40 light-years. This is a binary, consisting of a sunlike yellow star and a red dwarf, separated by about 1,000 AU. Astronomers have found five planets in the 55 Cancri system, ranging from several low-mass planets lying close in to a planet with nearly 4 times the mass of Jupiter, at a similar orbital distance to our largest planet.

In 2012, a research team announced that 55 Cancri e likely has a graphite-rich crust and probably a thick, diamond-rich layer underneath. So-called diamond planets were hypothesized years earlier, but 55 Cancri e became the first potential candidate (but there is widespread disagreement over these results). Although the planet's density is not much different from that of Earth, the host star is exceedingly rich in carbon, and given what they know about planet formation, astronomers believe the planet, at its distance, has a layer of diamonds kilometers thick below its surface. The planet has some 8 times the mass of Earth, and orbits its star in just 18 hours at a distance of only 2.4 million kilometers. At such a close distance, the planet warms to a searing 1750 °C (3200 °F), keeping those subsurface diamonds rather toasty.

The very first exoplanet ever found, around the pulsar PSR B1257+12, also has an odd tale to tell. In 1992, when Aleksander Wolszczan and Dale Frail made their discovery, they were studying the neutron star left behind when its progenitor massive star exploded. The remnant is as dense as an atomic nucleus and rotates once every 6.22 milliseconds. The radio pulses emitted from the

pulsar are extremely regular, and it was minor irregularities noticed by Wolszczan and Frail that led to the exoplanet discovery. The irregularities indicated the
presence of three planets in the system, located in the constellation Virgo in our
sky, at a distance of 1,000 light-years. Incredibly, this ancient, 3-billion-year-old
system would have destroyed any existing planets when the progenitor star
exploded. The pulsar planets discovered in 1992 are "second-generation
planets" that must have formed more recently from debris collapsing down
after the supernova explosion, or from material captured by a companion star.
Or perhaps they were stripped of their atmospheres.

Many of the earliest exoplanets to be discovered were hot Jupiters, and no
planet makes for a better example of one than WASP-12b, a planet discovered
by the SuperWASP planetary transit survey in 2008. This world orbits the star
WASP-12, a 12th magnitude sun in Auriga. The host star is a sunlike star that
lies at a distance of 870 light-years. The first exoplanet discovered orbiting a
main sequence star, 51 Pegasi, back in 1995, was a similar planet. But WASP-12b
is even more extreme. This planet has a mass some 450 times that of Earth, and
lies a mere 3.4 million kilometers from its star – the planet whips around the sun
once every 1.09 days. Because of its close distance and the gravitational tug
from the star, the planet is likely egg-shaped, and will eventually be torn apart
and consumed by the star. Observations made with the Hubble Space Telescope
bear out this conclusion, and mark the first time a planet has been observed in
the process of being torn asunder by its star's gravity. Moreover, recent studies
suggest the planet is exceedingly rich in carbon, but rather than in subsurface
diamonds, with WASP-12b the carbon is believed to be largely atmospheric, as
it is a gas giant.

On Earth, we think of our solar system as being unthinkably old at 4.6 billion
years. But an exoplanet circling a very faint star in the constellation Scorpius,
PSR B1620–26 b, at the great distance of 12,400 light-years, is one of the oldest
known planets – some 12.7 billion years old. Nicknamed "Methuselah," this
planet was discovered in 1993 by the pulsar timing technique. It orbits the
binary star PSR B1620, which consists of a pulsar and a white dwarf companion.
The pulsar is a neutron star that rotates 100 times per second, and the white
dwarf has a mass about one-third that of the Sun. The planet's mass is about
2.5 times that of Jupiter, and it orbits some 23 AU from the stars, which are
separated by the Earth–Sun distance, 1 AU.

Amazingly, the binary and its attendant planet lie just outside the core of the
well-known and rich globular star cluster M4, a favorite target for backyard
observers. PSR B1620–26 b is the first planet detected within a globular star
cluster. This system may have formed in stages. First, the massive star initially
belonged to a binary system. The larger star exploded, leaving behind a neutron
star and a small companion. And the planet formed around an unrelated star.
Then, astronomers say, the two systems collided about 2 billion years ago. The
neutron star's original companion was ejected, leaving the planet in a wide orbit
around the binary. When the sunlike star became a red giant a half billion years

ago, material swirled onto the neutron star, spinning it up into becoming a millisecond pulsar. This formed the weird outcome we can observe today.

Most exoplanets found thus far have been detected indirectly, and most of them orbit their stars closely and with relatively short periods. It takes time for astronomers to observe these planets, and the two dominant methods thus far – the Doppler technique and the transit technique – reveal close-in worlds more easily than planets lying farther afield. But an object imaged directly with the Hubble Space Telescope in 2008, and confirmed in 2012, is a rare detection of a planetary world encircling a well-known, bright star. (Some astronomers believe the Fomalhaut discovery may represent a knot of dust, however, rather than a planet. They refer to HR 8799, a young, main sequence star 130 light-years away in Pegasus, as a better direct-imaging planetary system. This A-type star has a debris disk and at least four massive planets, detected directly.)

If it is planetary, Fomalhaut b is a world orbiting the brightest star in the southern constellation Piscis Austrinus, Fomalhaut, a first-magnitude A-type star only 25 light-years away. Such a planet was suspected as early as 2005, when astronomers studied the massive dust ring that surrounds Fomalhaut – debris from the formation of the solar system. The ring is not exactly centered on the star, leading to the suspicion that a massive planet was also involved. The supposed planet probably has a mass no greater than twice that of Jupiter, and orbits the star at about 115 AU, just inside the inner edge of the large dust ring. It is quite possible that other planets also exist in the system; they would help to explain some of the dynamics of where the observed planet lies.

One of the most spectacular exoplanet discoveries came along in 2011, when astronomers announced their discovery of Kepler-70b. This planet orbits the subdwarf star Kepler-70, a 15th-magnitude sun lying 3,900 light-years away in the constellation Cygnus. This remarkable planet is the closest to any star known, orbiting at the incredibly small distance of 900,000 kilometers, only 1/65th the distance between Mercury and the Sun. Consequently, the planet has the fastest known orbital velocity, some 980,000 kilometers per hour, and it has the smallest known mass of any exoplanet, at just 44 percent of Earth's mass. Orbiting so close to its star, Kepler-70b is also the hottest known exoplanet, radiating at 6930 °C (12,500 °F).

But the story gets even stranger. Kepler-70b has a sister planet, Kepler-70c, that is slightly larger, cooler, and more distant from the star. Kepler-70 spent several billion years on the main sequence and has evolved away from it, becoming an aged, red giant, until helium burning commenced in its core some 18 million years ago. This shrank the star to its current size. When the star was in its red giant phase, the two planets would have been inside the star's diameter. Astronomers believe they were once gas giants, but as the star's outer layers swelled, it evaporated the planetary atmospheres, leaving behind dense cores – all that remains of the two planets.

The quest for detecting planets around other stars is all about the excitement of knowing how common life might be in the cosmos. Do millions of

Figure 8.5 NASA's Far Ultraviolet Spectroscopic Explorer spacecraft found a dusty disk around the young star Beta Pictoris – a system less than 20 million years old that could already be home to planets.
NASA/FUSE/Lynette Cook

civilizations exist in the galaxy? Several? Just one? This largest of all philosophical questions drives the quest to categorize worlds. And, of course, the star system nearest to us has always loomed large in that discussion. The closest star system to us, Alpha-Proxima Centauri, lies a little more than 4 light-years away. Astronomers have always wanted to find a planet more or less like Earth that orbits a sunlike star – an analog to our own solar system – and in 2012, this occurred. A group of European astronomers announced the discovery of Alpha Centauri Bb, a planet orbiting the "B" component of the binary star that composes Alpha Centauri. Although some controversy lingers over the existence of this planet, the data are promising. The planet is some 10 percent heavier than Earth and lies 4.37 light-years away, encircling the star Alpha Centauri B, a K-type star shining at 1st magnitude, with a mass 93 percent that of the Sun. (Alpha Centauri A is a G-type star like the Sun, with 10 percent more mass than our star; Alpha Centauri C, also called Proxima Centauri, is a reddish M-type star in a wide orbit about the binary, and so lies closest to us, at a distance of 4.24 light-years.)

To the disappointment of many astronomers, Alpha Centauri Bb does not lie in the star's habitable zone. It lies a mere 6.3 million kilometers from its star, so the planet's surface temperature is likely a broiling 1200 °C (2200 °F). The two main stars, Alpha Centauri A and B, orbit a common center of mass once every

80 years at an average distance of 17.6 AU. This establishes a very intriguing scenario for a complex system of suns and at least one planet in the very nearest star system to us – although, again, many astronomers caution that this result needs more observations for complete conformation. But it does show us that quite possibly, we have an intriguing set of worlds nearest us in the galaxy, and among the two systems – Alpha Centauri and our own – we are definitely in the best place for the existence of a stable life.

Chapter 9
The Milky Way as barred spiral

When I first got into astronomy in the mid-1970s, one of the things that grabbed me most significantly was observing night after night in a cornfield at the back of our house, adjacent to our suburban subdivision. I had quite a dark sky outside our little town in southwestern Ohio and could endlessly explore treasures of the night sky, at first with naked eyes and a pair of binoculars. On summer nights, the long arch of the Milky Way, stretching from Cassiopeia in the north through Cygnus and all the way down into Scorpius and Sagittarius in the south, was simply amazing. Filled with glistening stars and pockmarked by myriad rifts of dark nebulae, it was mesmerizing, night after night.

During those first few weeks of my astronomy hobby, I didn't really grasp entirely what I was looking at. And for most of the history of astronomy, no one did. Only in the early 1920s did Edwin Hubble make a breakthrough discovery that led to understanding the true nature of galaxies, which were thought to be "spiral nebulae" and possibly within our own galaxy, for decades beforehand. Aided by discoveries by Lowell Observatory astronomer Vesto M. Slipher (1875–1969), Hubble and others deciphered the cosmic distance scale, at least to a first approximation, and by the mid-1920s astronomers and the informed public understood that we live in a cosmos filled with numerous galaxies and that the Milky Way is just one of them.

Classifying the types of galaxies, however, has been a long and somewhat arduous process. Hubble's original classification scheme, derived following numerous observations and proposed in 1936, consists of a "tuning fork" diagram showing elliptical galaxies on the "handle" branching out into various types of spiral galaxies and barred spiral galaxies on the "tines" of the fork. The galaxies are classified not only by their rough shapes, ellipticals being big spheres of stars, some slightly flattened and some purely spherical, and spirals showing spiral arms, barred spirals being spirals with a prominent bar of material running through their centers, from which the spiral arms originate. The original Hubble classification did not include other types of galaxies such as lenticular (lens-shaped) galaxies, irregular galaxies, and peculiar galaxies,

Figure 9.1 The center of the Milky Way in Sagittarius, Scorpius, and environs imaged by the author, who employed his camera from the very dark sky outside San Pedro de Atacama, Chile, in the spring of 2014. Eicher used a Canon 6D camera, a Sigma 12mm lens at f/3.5, and an approximately 20-second exposure, handheld, bracing the camera between rocks on the roof of a hut. This is a testament to the darkness of the Chilean sky!
David Eicher

which were not well categorized in the mid-1930s. Later, Hubble's student Allan Sandage (1926–2010) expanded and refined the categorization.

But a more definitive classification scheme for galaxies came along with the work of Gérard de Vaucouleurs (1918–1995), a French astronomer who spent most of his career at the University of Texas in Austin. In 1959, while at Harvard University, de Vaucouleurs began working on an improved system that reflected the increased complexity of the types of galaxies astronomers had observed over the intervening 20 years since Hubble's initial scheme. The scheme proposed by de Vaucouleurs, still in use as the standard, retains the basic types outlined by Hubble: ellipticals, spirals, and barred spirals – and lenticulars added by Sandage. But de Vaucouleurs also added an elaborate system noting various prominent features of many galaxies to further classify bars, rings, and spiral arms, turning Hubble's two-dimensional graphic into a more complex, information-packed, "three-dimensional" representation of galaxy types that is sometimes nicknamed the "lemon slice."

When it comes to classifying and understanding galaxies, we have a rich heritage of observations. The universe contains some 100 billion galaxies, most astronomers estimate, and we see numerous types of galaxies in the universe at a huge range of distances. That means we are seeing galaxies in many stages of evolution, from mature galaxies relatively near us in space to very young galaxies at great distances as they appeared in the early days of the universe.

Categorizing those galaxies has been a pretty straightforward exercise as our understanding has increased, from the discovery of "spiral nebulae" by the Irish observer William Parsons, Third Earl of Rosse (1800–1867), who observed and sketched the Whirlpool Galaxy and others in 1845, to Hubble, to Sandage, to de Vaucouleurs. But our own galaxy gives us a special challenge. That arching band of light in the sky is the unresolved light from billions of stars in the disk of the Milky Way Galaxy, and it's particularly challenging figuring out the basic information about a galaxy when you are inside it.

The term Milky Way itself comes from the Latin *via lactea*, referring to the milky appearance of the band of light in the sky, which in turn originated from the Greek *galaxias kýklos*, meaning "milky circle." It was our old friend Galileo Galilei who first resolved stars in the Milky Way and had the first glimmer of understanding it, late in the year 1609. The enormous challenge of deciphering the physical structure of the Milky Way meant that nearly 400 years had to pass before astronomers really understood its nature with any degree of high precision. Although mapping the structure of the Milky Way is an endeavor that began in the 1950s and solid evidence for the central bar of the Milky Way emerged in the early 1990s, progress toward the ultimate goal of a complete map of the Milky Way has only recently started to accelerate. This is due in large part to large-scale infrared surveys of the galactic disk and bar like GLIMPSE, a survey project using the Spitzer Space Telescope (more on that later in this chapter).

Before we get into the revelation of the Milky Way as a barred spiral, however, let's examine the basics of what it contains. The galaxy has several main components, which include the most prominent part, a thin disk of stars, which is roughly circular and contains dark patches of dust and dense gas that are plainly visible in a dark sky with the naked eye. Distances relating to major features in the galaxy are now reasonably well known. The stellar disk extends out to about 44,000 light-years on either side of the nucleus; beyond this distance, it continues, but the density of stars drops off significantly and the disk begins to flare. The Milky Way's Central ("Bulgy") Bar extends some 11,400 light-years on either side of the nucleus. The outer limits of the gas and star formation in the Milky Way extend to about 72,000 light-years from the nucleus.

In 2000, a group of astronomers at the Institute for Astrophysics in Tenerife, in the Canary Islands, published confirmation of the galaxy's so-called Long Bar, which extends out some 14,000 light-years on either side of the nucleus. This bar was confirmed by the GLIMPSE survey in 2005. More on that to come. The best estimate of the Sun's distance from the galactic center is 27,100 light-years.

So, to summarize, the best measurements of the diameters of features in the Milky Way Galaxy follow: the Central, "Bulgy" Bar is 22,820 light-years across; the Long Bar is 28,700 light-years across; the position of the Sun is

Figure 9.2 The Whirlpool Galaxy and its companion, NGC 5195, as captured with the Hubble Space Telescope, and a strikingly similar drawing of the same "spiral nebula" made in the 1840s by William Parsons, the 3rd Earl of Rosse, with his large telescope in Ireland.
NASA, ESA, S. Beckwith (STScI), and The Hubble Heritage Team (STScI/AURA)

Figure 9.3 This annotated roadmap of the Milky Way highlights our names for its many arms. Our Sun sits in the Orion Spur, packed just in from the much larger Perseus Arm. NASA/JPL-Caltech/R. Hurt (SSC-Caltech)

27,100 light-years away from the galactic center; the boundary of the main stellar disk is defined as about 88,000 light-years across; and the outer limits of gas and star formation extend over a diameter of some 143,000 light-years.

The thin disk contains 90 percent of the stars in our galaxy's disk, and all of the young, massive stars that have been born in open star clusters. Most of the material in this thin disk lies within a "platter" some 1,500 light-years thick. The remaining stars, scattered somewhat outside the realm of the thin disk, form a thick disk of material that is some 3,000 light-years in thickness. The stars in the thick disk formed earlier in the galaxy's history than those of the thin disk. And the gas and dust in the disk of the Milky Way is chiefly arranged in a thin layer, most of it less than 500 light-years from the center of the disk.

Stars and gas orbit the center of the Milky Way, and the gravitational effects eventually are stupendous. The Sun was born in an open star cluster some 4.6 billion years ago, but our cluster mates are long since dispersed, leaving the Sun as a solitary star, the companions to our star sent in other orbits as they were tugged by the galactic center and/or randomly passing stars. We are moving around the center of the galaxy at a velocity of about 240 kilometers per second, and one orbit around the galaxy takes about 220 million years. Thus, the Sun has orbited the galactic center a grand total of 20 times. The total luminosity of the disk is about 20 billion times that of the Sun, and the mass is about 60 billion solar masses.

The galaxy also contains a central bulge and bar, which extends some 28,700 light-years across. Within the galaxy's center lies a supermassive black hole, Sagittarius A*, which contains some 4.1 million solar masses. The galaxy's bulge contains some 5 billion times the luminosity of the Sun and some 20 billion solar masses. The halo, which contains metal-poor globular star clusters, clouds of neutral hydrogen gas, and large amounts of dark matter, extends to more than 200,000 light-years on both sides of the galactic center. Stars in the halo contribute only a small amount of the galaxy's mass, about 1 billion solar masses. Using orbital calculations, astronomers find that most of the galaxy's mass must lie beyond 30,000 light-years from the center, where relatively few stars lie. Thus, the amount of dark matter in the galaxy's outer dark halo must be enormous, approximately 90 percent of the total mass of the galaxy.

Figure 9.4 The Milky Way's center is home to a supermassive black hole packing millions of Suns worth of material, as well as hundreds of white dwarfs, neutron stars, and black holes surrounded by a fog of gas. Sagittarius A*, as the monstrous black hole is known to astronomers, can be seen at the center in this Chandra spacecraft mosaic. The object glows bright in both x-rays (as shown here) and radio wavelengths, but its light vanishes in visible wavelengths because of the dust between Earth and our galaxy's center.
NASA/UMass/D. Wang et al.

Even experienced astronomy enthusiasts who are quite knowledgeable often underestimate how much empty space there is in space. In the vicinity of the Sun, astronomers find on average about one star in every 350 cubic light-years. Most interstellar space, therefore, contains gas and dust but few stars. Radiation from this interstellar medium arrives in our solar system from ionized gas, from molecules, and from neutral atoms.

Some 1 percent of the interstellar medium consists of dust particles, which are mainly silicate particles and carbon-rich grains, and most are exceedingly tiny – smaller than 1 micron, about the size of particles in cigarette smoke. Vast clouds of them are often visible as dark nebulae because they obscure light from stars and other objects lying beyond them.

The Sun's position in the galaxy gives us an incredible station from which to observe the daily happenings in a reasonably large barred spiral. Distances to astronomical objects, even the stars nearest the Sun, are hard to measure. Trigonometric parallax allows measuring the distances to the nearest stars. More distant stars can then be compared to stars close by, and assuming the stars with similar spectra have similar properties, distances for them can be derived by comparing apparent brightnesses with assumed absolute brightnesses. Trigonometric parallax is the basis for nearby stars, however, up to about a thousand light-years. As Earth orbits the Sun, our reference frame changes slightly and we can observe how stars move relative to distant objects. During the course of a year, this effect allows measuring stars near to us pretty accurately, as the Hipparcos satellite did from 1989 to 1993, mapping out 120,000 stars closest to us in the galaxy with very high precision. The currently operating Gaia mission will extend this to a billion stars.

But most observable stars do not have parallaxes, demonstrating that the galaxy is extremely large. To determine distances to most stars, we need to rely on the cosmic distance ladder, measuring their distances relative to stars that are close enough to show parallaxes. Sometimes velocities of objects can be used to derive distances, and this helps to map out the distribution of stars and star clusters in the Milky Way. Distances from motions are important – one current best estimate of the size of the galaxy itself comes from the velocities of stars orbiting the Milky Way's central black hole. The Sun's distance from the black hole and therefore the galactic center, derived from these motions, is 25,000 light-years. A March 2014 measurement, however, made at high precision with the Very Long Baseline Array of radio telescopes concludes a distance of 27,100 light-years.

Spectroscopic parallax is also an important tool; that is, examining the width and depth of a star's spectral lines, which betray its absolute luminosity, and then using the apparent brightness to derive a distance. This technique works better with some types of stars than with others, but it is key in estimating some of the star-forming history of the galaxy, at least as judged by our solar neighborhood. Assuming our neighborhood is typical, the Milky Way has made stars at a rate of three to five solar masses per year in order to have built the stellar disk. And the disk contains between 5 and 10 billion solar masses of cool gas, so it is possible to sustain this rate for several billion years into the future.

Further measurements of this type allow for creating a database of scale heights – the thickness of each component of the galaxy – and orbital velocities about the galactic center. So astronomers believe carbon monoxide in the Sun's region extends to about 400 light-years in thickness; layers of cold and warm neutral hydrogen gas extend to 1,300 light-years in thickness; ionized hydrogen gas extends to 3,300 light-years in thickness; and the layer of stars in the disk extends to 1,000 light-years (the thin stellar disk) and 3,300 light-years (the thick stellar disk).

Stars within the thin stellar disk orbit at velocities around 40 kilometers per second; stars within the thick stellar disk orbit at some 65 kilometers per second; and objects in the halo orbit at around 100 kilometers per second, at least those at about 75,000 light-years' distance from the galactic center.

The thick disk is much less densely populated than the thin disk, with only 10 to 30 percent the amount of stars, gas, and dust. It contains no O, B, or A stars, so the thick disk is older than the thin disk, probably older than 3 billion years. (The thin disk is, of course, still forming stars.) The thick disk may in fact be the remnant of a precursor thin disk that existed in the early history of the Milky Way. We know that galaxies form in the universe by merging from many smaller protogalaxies that ultimately create larger ones – witness the many small, bluish blobs of early galaxies in imagery like the Hubble Ultra Deep Field. The Milky Way may have formed from something like 100 tiny galaxies that gravitationally came together to make the structure we have today. Mergers could have shaken up the earlier thick disk and distorted it, randomizing orbits, and depositing a plane of gas into the galaxy that formed the stars in the thin disk we now see.

Stars like the Sun in the thin disk are born in star clusters and stellar associations, and the Sun lies within an association of stars called Gould's Belt, a ring or disk of bright young stars. Within a radius of about 1,500 light-years, young stars are not inside the plane of the thin disk, but are rather arranged in a line slightly tilted, by about 20°, to the thin disk. In another few galactic orbits, however, the stars within Gould's Belt will disperse completely into the thin disk. They will still follow the same velocities and directions they have now, in large part, and will then be called a moving group. The most interesting and nearest moving group to us now in the Ursa Major Moving Group, which contains many of the bright stars of the Big Dipper asterism, moving along in space together, aligned by the same velocities and direction through the galaxy from their stellar birthplace.

Stars are born in clusters, and we can learn a lot about the Milky Way by studying stars that are still in clusters. The famous Pleiades (M45) in Taurus, visible to the naked eye as a tiny dipper-shaped asterism, often called the "Seven Sisters" because of its seven stars visible to the naked eye, is a great example of a nearby open cluster. (Seven stars are visible hypothetically; you may only see six unless you have perfect vision and a perfect sky.) Open clusters are the norm when it comes to starbirth; they are sometimes called galactic clusters, a now somewhat outdated term. The Pleiades contains some 700 stars brighter than magnitude 17. Astronomers can learn a lot about the composition of star clusters by plotting their member stars on a color-magnitude diagram (CMD), recording the brightnesses of cluster members on the Y-axis and the colors or temperatures of the stars on the X-axis. By comparing CMDs of various clusters, they can compute ages and distances to the clusters. Through such computations, astronomers know the Pleiades is about 400 light-years distant, contains 800 solar masses, and is some 16 million years old.

Viewing open star clusters gives us a window back in time to see what the Sun's neighborhood must have been like 4 billion years ago, when Earth and the other planets were young. Their hundreds of stars are bound by gravity and they can also contain gas and dust in appreciable amounts. Because of the dust in the Milky Way's disk, however, we see only a small percentage of the open clusters that exist in our galaxy. Most open clusters are younger than 300 million years, and only a few are older than a billion years. Because most of the light in a cluster comes from its brightest stars, the integrated light from an open cluster gives astronomers a measure of its age. Star clusters redden over time as massive stars die and the cooler stars are left to shine on in greater numbers.

By contrast, globular star clusters, huge spheres of stars lying in the galaxy's halo, are composed of older, redder stars. The largest globular in the Milky Way is Omega Centauri, visible in the southern sky; it contains more than a million stars and is larger than the diameter of the Moon in our sky, despite the fact that it is some 15,000 light-years away. The Milky Way contains about 150 globular clusters; some larger elliptical galaxies hold enormous clouds of many hundreds of them. Omega Centauri's luminosity of a million suns is impressive compared with some smaller globulars, which have luminosities of only a few tens of thousands that of the Sun.

Globular clusters are really most impressive in their antiquity. Most in the Milky Way are at least several billion years old, and the CMDs of most globular clusters show no young stars at all. Another prominent globular cluster in the southern sky, 47 Tucanae, is older than 10 billion years. The most metal-poor globular clusters have ages of about 11 to 12 billion years – thus, the stars in some globular clusters, and by association the globular clusters themselves, formed before the main formation of the Milky Way itself took place, which is thought to have been about 9 billion years ago.

The belief today is that as galaxies like the Milky Way started to form, gas under conditions of extremely high density and pressure formed these richly populated star cities; the kinds of conditions that formed globular clusters did not exist except early in the evolutionary history of the galaxy. Once these extreme conditions were gone, globular clusters could no longer form.

Globular clusters are too far away to have their distances measured by parallax. Rather, astronomers compare their CMDs with theoretical models based on stellar evolution and adjust them to find a best fit in terms of the stars they contain, which then provides a distance estimate. Also, variable stars called RR Lyrae stars, pulsating variables with known intrinsic brightnesses, can be used to estimate their distances.

Globulars also give us some interesting information about how the Milky Way formed. Because stars are made from dense gas clouds, astronomers would expect the oldest stars to be located in the center of the galaxy, where the densest gas lies. But the oldest stars are in globulars, scattered far away from the galaxy's center, in the halo. How did this happen?

The answer goes back to the fact that galaxies grow by eating each other. The center of the galaxy, even its young disk, is an active conveyor belt of activity; old objects have fallen in and been consumed in mergers, creating waves of new star formation. The Sagittarius Dwarf Galaxy, a small satellite of the Milky Way, is now in the process of being eaten. Omega Centauri has such an odd range of metallicities in its stars – of elements heavier than hydrogen and helium – that astronomers suspect it may be all that's left of a now-dead dwarf galaxy. Within the next 5 billion years, the Magellanic Clouds will share the fate of the Sagittarius Dwarf, falling into the plane of our galaxy.

So globular clusters are a main constituent of the so-called metal-poor halo of the Milky Way. But the halo also contains roughly 100 times more metal-poor individual stars, outside of globular clusters, in weird, highly eccentric orbits around the galactic center. In our neighborhood of the galaxy, about 1 percent of the stars belong to the metal-poor halo, brought in temporarily to our vicinity as they swing past in strange orbits that will again carry them far away. Unlike normal stars in the disk, these stars are moving fast relative to the Sun. These stars can be close to us, but they can also be as distant as 300,000 light-years from the galactic center.

When we turn our gaze inward toward the center of the Milky Way, astronomers have always been stymied by copious amounts of dust in the disk. This means that infrared astronomy, which enables them to peer through dust clouds and on beyond, is the key to studying the Milky Way's bulge and center. Such studies show that the galaxy's flattened central bulge contains about 20 percent of its light. The bulge is peanut-shaped, and most of its light is concentrated within the inner 3,000 light-years. The apparent peanut shape results from the galaxy's central bar, which extends about 12,000 light-years from the center.

For some time, astronomers wondered whether the bulge was just an increasingly dense part of the halo, but the bulge is very real and a separate physical entity. Stars in the galactic bulge orbit the center in the same direction as disk stars. The area close to the galactic center is packed with extremely dense gas and young stars. An enormous star cluster, Sagittarius B2, lies about 450 light-years away from the center of the galaxy and is producing new stars at a huge rate. In the area 100 to 150 light-years from the galactic center, the Arches and Quintuplet clusters are each a million times more luminous than the Sun. Very close to the galactic center, astronomers find huge, hot, dense molecular clouds, and stars in the nucleus of the galaxy.

We cannot see the center of the galaxy in optical wavelengths. Too much material lies between the galactic center and our solar system, 27,000 light-years out in the suburbs. But infrared observations show the stellar nucleus is more or less like a massive globular cluster, holding 30 million solar masses within a diameter of just 1.3 light-years. In this densest region of the galaxy, some 30 massive stars have formed in the last 7 million years.

These massive young stars reach to within a tenth of a light-year of the galaxy's central radio source. Maps made in radio wavelengths of the inner galaxy show filaments tens of light-years long reaching up above and below the galactic plane.

These are probably magnetized (like the coronal arches on the Sun) and are indicative of extreme conditions – high pressures, densities, and magnetic fields – in the environs of the center of the galaxy. Some may be involved with the galaxy's central black hole, Sagittarius A*, which contains 4.1 million solar masses. This compact source, which astronomers deduce is a black hole because of its extraordinarily high calculated density, must measure only about 20 million kilometers across. The energy output of this black hole is several tens of thousands greater than the Sun's, but still falls far short of the monster black holes in the centers of huge active galaxies.

Arranged between the stars of the Milky Way lie blankets of interstellar gas. Although the gas in our galaxy represents a mere 10 percent as much mass as our galaxy's stars, the gas component contributes significantly to how our galaxy works. The presence of gas allows stars to actively form, keeping our galaxy classified as a barred spiral; without the gas, it would be a lenticular galaxy that would lack its spiral arms. Due to gravity, this interstellar gas eventually gets recycled into new stars as old stars slowly return gas into the interstellar medium.

The densest parts of the Milky Way's gas, called giant molecular clouds, lie along spiral arms and can stretch over 60 light-years and contain several hundred thousand solar masses of material. These clouds are typically surrounded by less dense clouds of cool atomic hydrogen. This gas often is mixed in character, temperature, and density. At present, the Sun is moving through a cloud of warm gas, about 50 percent of which is ionized, and which spans about 3 light-years across. This cloud lies within an expanding so-called Local Bubble, which measures about 300 light-years across.

The layers of gas residing in the Milky Way are also acted on by magnetic forces. When a supernova explodes into a cloud of gas, the shockwave accelerates protons and other particles to extremely fast speeds, creating cosmic rays. Some of these cosmic rays bombard us nearly constantly on Earth. Like stars, molecular clouds of gas orbit the galactic center, although typically at pretty slow speeds. They are arranged in all manner of orbits.

Figure 9.5 The sky burns with stars, but the energetic source that contains the black hole at the Milky Way's center, Sagittarius A*, is invisible in this image, captured by an infrared camera on the Hubble Space Telescope.
NASA, ESA, and G. Brammer

So you can see that the Milky Way contains a huge variety of objects, just from the few we have visited. How was it, then, that astronomers recently came to reassess the shape and architecture of our galaxy, determining that it is indeed a barred spiral?

As we have seen, for decades astronomers believed the Milky Way was a simple spiral galaxy, like the Andromeda Galaxy but on a somewhat smaller scale. Untangling a more sophisticated and accurate view of our galaxy required a huge set of observations spanning a variety of wavelengths across the electromagnetic spectrum.

The first evidence for the spiral structure in our galaxy dates to 1951, when American astronomer William W. Morgan (1906–1994) of Yerkes Observatory unveiled the first map of HII regions and hot O and B stars in the vicinity of the Sun. These appeared to show segments of the nearest spiral arms. But because our view of the inner part of the Milky Way is obscured by dust in the optical wavelengths, most of the effort to map the Milky Way moved to using 21 centimeter radio waves emitted by hydrogen gas clouds.

By measuring the Doppler shifts of these gas clouds, astronomers could identify many, but not all, bands of gas clouds that they interpreted as the Milky Way's spiral arms. This was a long-term projection initiated in 1952 and led by the Dutch astronomer Jan H. Oort. But due to several difficulties in converting Doppler shifts into distances, astronomers could not reach agreement on a reliable map of the galaxy's gas, and by the early 1980s, this technique seemed to have run aground.

By the early 1990s, astronomers marked the discovery of a prominent bar in the Milky Way, a diagonal band of stars running through the center of the Milky Way. Although peculiarities in the galaxy's gas flow had convinced some astronomers that such a bar might exist, the availability of low-resolution infrared images, notably from the COBE-DIRBE instrument, showed a vertically extended bar in the center of the galaxy. (The Diffuse Infrared Background Experiment, DIRBE, was an experiment conducted with the COBE spacecraft to survey the infrared sky.) Because infrared light penetrated the dust more easily, astronomers could see the bar directly for the first time, even though the bar was not resolved into stars.

More recently, also in the infrared part of the spectrum, a project overseen by astronomers Ed Churchwell of the University of Wisconsin and Robert Benjamin of the University of Wisconsin-Whitewater, has led to new advances in our view of the galaxy in the first decade of the twenty-first century. The opportunity arose in 2003 when NASA launched the Spitzer Space Telescope, one of the great orbiting observatories, and the most advanced infrared telescope ever made. One of the Spitzer projects was called the Galactic Legacy Infrared Mid-Plane Survey Extraordinaire (GLIMPSE).

GLIMPSE used the Spitzer Space Telescope to survey more than 220 million stars and most of the galaxy's gas clouds, and produced a new understanding of the Milky Way's structure beginning with its first major results in 2005. GLIMPSE focused on the distribution of the stars and infrared-bright star forming regions, creating a composite map of the inner galaxy, its molecular ring of gas, the number and location and spiral arms, and its central bar.

This would not have been possible with the complexity of Spitzer's infrared instruments. The GLIMPSE team took advantage of the telescope's Infrared Array Camera to survey a huge strip of the galaxy, $1°$ above and below the galactic plane. More than 90 percent of the sources initially catalogued were red giant stars, which are luminous and therefore can be seen from large distances. The mid-infrared is not perfect. Dust still blocks some objects from beyond, but the GLIMPSE team was able to record enough data to produce an enormously improved map of the Milky Way.

By counting stars at equal angles relative to the galactic center, the GLIMPSE team confirmed the existence of a central bar. The bar contains a large population of stars known as red giant clump stars, which shine with a fixed

Figure 9.6 A laser takes aim at the wide band of gas and stars occupying the Milky Way's galactic center as stars trail overhead. Yepun, one of the four 8.2-meter telescopes of the European Southern Observatory's Very Large Telescope in Paranal, Chile, stands in the foreground.
Dave Jones/ESO

luminosity and could be used as "standard candles" to measure the bar's length and orientation, finding that the bar extended out to a much larger radius than most astronomers had previously believed.

Secondly, the GLIMPSE team produced a refined picture of our galaxy's spiral arms. By counting stars, they were able to study closely two areas where huge numbers of stars were expected, along the line of sight of two spiral arms. One of the arms, the Scutum (or Scutum-Crux or Scutum-Centaurus) Arm, shows a terrific enhancement of stars as expected from optical and radio studies. Another region rich in stars, however, the Sagittarius (or Sagittarius-Carina) Arm, showed no strong enhancement in stars from the GLIMPSE star counts. The unusual history of star formation in the Sagittarius Arm will no doubt be a topic of ongoing studies.

The picture supported by GLIMPSE, however, was that the Milky Way is indeed a barred spiral, probably classified as an SBc barred spiral. Originating at the ends of the galaxy's central bar are several prominent spiral arms. They include the Perseus Arm, one of the galaxy's two major spiral arms, which lies some 6,400 light-years from the Sun. Associated with the Perseus Arm are the Near 3 kpc Arm and the Far 3 kpc Arm, smaller, attached components. The Scutum Arm is another major arm, extending from the other end of the bar from the Perseus Arm. The Perseus and Scutum Arms are suspected to be the major arms of the galaxy – they contain huge numbers of old stars as well as plentiful gas.

Also originating near the bar's terminus is the Norma and Outer Arm or Arms, along with a faint extension. The Carina-Sagittarius Arm is also a somewhat more minor arm. Another minor arm, or spur, is the Orion Arm, which contains the Sun and Earth. It stretches some 3,500 light-years in width and is some 10,000 light-years long. The Orion Arm lies between the Carina-Sagittarius Arm and the Perseus Arm. Thus, we see from our vantage point the constellations Carina and Sagittarius when we look generally toward the center of the galaxy, and the constellation Perseus and surrounding environs when we look away from the center of our galaxy.

In 2013, astronomers at the National Radio Astronomy Observatory announced a refinement of the status of the Orion Arm, using the National Science Foundation's Very Long Baseline Array radio telescope. They suggested that the Orion Arm, thought to be a minor spur, should be considered an upgraded neighborhood that is actually a more significant structure. By looking at masers in the local galaxy, they could examine water and methanol molecules and derive more accurate distances to objects in the Orion Arm, suggesting it is not merely a spur but larger than previously believed.

These new results underscore how tentative our understanding of the structure of the Milky Way is. Four hundred years after Galileo first resolved stars in the galaxy, and nearly a century after Hubble and Slipher set the galactic distance scale, we still have much to learn. But how the last decade has provided an explosion in the better understanding the system of stars, gas, and dust in which we live!

Chapter 10
Here comes Milkomeda

Not only do we now understand the basic structure of our galaxy well, but also recent years have unveiled the fate of the Milky Way. While the universe is expanding and on large scales, objects are moving apart from each other, on smaller scales gravity and local motions can bring objects together. Astronomers have seen evidence of interactions, collisions, and mergers of galaxies in numerous dense groups and clusters of galaxies. In fact, galaxy mergers are common mainly near the dense center of a cluster, contributing to the growth of the most massive galaxy (called a cD galaxy), which lies at the cluster's heart. In such groups, driven by gravity, galaxies eat each other. Larger galaxies can become larger and larger over time, giving these collections of stars, gas, and dust the ability to grow into massive bodies far more massive than their original mass.

We do not live in a dense cluster of galaxies like the Hercules Cluster, the Virgo Cluster, or the Coma Cluster. But even in our loose and small group of galaxies we have experienced – and will experience – mergers. To understand why, you first need to understand the dynamics of our own galaxy group.

Just more than a decade after his discovery of a Cepheid variable star in the Andromeda "nebula," since known as the Andromeda Galaxy (M31), Edwin P. Hubble defined the nearest galaxies to us as the Local Group of galaxies, explained in his 1936 work *The Realm of the Nebulae*. The first nine members Hubble identified in our little group were the Milky Way, the Large Magellanic Cloud, the Small Magellanic Cloud, the Andromeda Galaxy, M32, NGC 205, the Pinwheel Galaxy (M33), Barnard's Galaxy (NGC 6822), and IC 1613. He listed as possible members IC 10, IC 342, and NGC 6946. In 1936, Hubble knew of nine members and perhaps as many as twelve.

Nearly 80 years later, we know a vast amount more about the Local Group of galaxies. The Local Group contains at least 50 galaxies, and astronomers are discovering tiny Local Group galaxies every now and then. There may be more than 100 galaxies in the group, spanning a sphere of space some 10 million light-years across. Observers keep finding more faint dwarf galaxies in the Milky Way's halo, and only a fraction of the sky has been surveyed to the maximum depth possible. The group's center of mass lies between the two dominant galaxies, the Andromeda Galaxy and the Milky Way.

In 1959, the German–English astronomer Franz Kahn (1926–1998) and Dutch astronomer Lodewijk Woltjers (1930–) demonstrated the so-called

Figure 10.1a The Andromeda Galaxy approaches for its imminent merger with the Milky Way in this illustration of an earthbound view 2 billion years in the future.
NASA, ESA, Z. Levay and R. van der Marel, STScI; T. Hallas; and A. Mellinger

Figure 10.1b Nearly 3 billion years in the future, the Andromeda Galaxy looms large in Earth's night sky.
NASA, ESA, Z. Levay and R. van der Marel, STScI; T. Hallas; and A. Mellinger

Figure 10.1c Star formation runs rampant in Earth's night sky, as the two galaxies merge into one massive collection of stars.
NASA, ESA, Z. Levay and R. van der Marel, STScI; T. Hallas; and A. Mellinger

timing argument, in which they proposed that Andromeda and the Milky Way formed within close proximity to each other in the relatively early universe before being pulled apart by cosmological expansion. By now, however, gravitational forces have taken over and the two galaxies have reversed direction relative to each other. The two galaxies have by now orbited relative to each other over more than half a period. The motion of the two galaxies is nearly radial; a full period would correspond to the time they merge, and half a period corresponds to the maximum separation between the two. By employing those age-old wonders – the laws of motion devised by Johannes Kepler – astronomers can understand some fundamental properties of the Local Group.

The timing argument of Kahn and Woltjers led to estimating the mass of the Local Group at more than 3 trillion times the mass of the Sun. Additionally, with respect to the Milky Way and Andromeda galaxies, they found the semi-major axis of the orbit – half of its long axis – at less than 1.9 million light-years. Curiously, they also found that a close encounter between the galaxies would occur somewhat more than 4 billion years from now.

Before getting into the implications of this, however, consider the Andromeda Galaxy itself. (We pretty thoroughly went through the structure of the Milky Way in the previous chapter.) The Andromeda Galaxy lies about 2.5 million light-years away, so the light reaching us now, the photons you see in a telescope's eyepiece, departed the galaxy while *Australopithecus africanus* roamed the plains of Africa, slightly before the emergence of *Homo habilus*, the first species to precede us in our genus.

The nearest spiral galaxy to us, M31, is classified as an SA(s)b spiral in the system devised by the French astronomer Gérard de Vaucouleurs. This means it is a spiral galaxy without a central bar (SA), without a ring (s), and with moderately tightly wound arms (b). In 2005, however, astronomers using data from the Two Micron All-Sky Survey (2MASS) project published results suggesting the galaxy has a box-shaped central bulge characteristic of a barred spiral galaxy. Suspicions over this had existed for many years, but the galaxy's high inclination to our line of sight and the obscuring effects of dust make confirming such a structure very difficult.

Strangely, the galaxy's disk is warped significantly in the shape of an S. This may have resulted from a close encounter with the number three galaxy in the Local Group, the Pinwheel Galaxy, that occurred some 2 to 4 billion years ago – or by an encounter with M32. Alternatively, the warping may be caused by some of the galaxy's satellites. The Andromeda Galaxy has a following of at least 14 dwarf galaxies that orbit the large spiral. A considerable number of additional objects are suspected dwarf companions.

The two prominent dwarfs visible in backyard telescopes, in the same low-power field as M31, are M32 (NGC 221), a bright, spherical cE2 galaxy, and NGC 205 (sometimes referred to as M110), an elongated, E5-pec galaxy of lower surface brightness. Other Andromeda satellite galaxies include NGC 147 and NGC 185 (in Cassiopeia, and visible in backyard telescopes), the Cassiopeia and Pegasus Dwarf Galaxies, and a slew of confirmed and suspected objects identified as Andromeda I, Andromeda II, and so on.

The Andromeda Galaxy has figuratively grown in recent years, thanks to a higher precision of astronomical research. In 2005, astronomers at the Keck Observatory in Hawaii studied lanes of stars emanating outward from the galaxy's disk to demonstrate they were actually a part of the disk, and therefore it is significantly larger than was previously thought. This extended stellar disk of M31 is substantially larger than the Milky Way's. The most heavily populated portions of the two galaxies are much smaller, however. The so-called disk scale radius of M31 is 17,000 light-years (and corresponding diameter 34,000 light-years), which defines the area of the disk in which the number of stars drops exponentially with distance from the center. The scale radius represents the distance of the exponential drop. The Milky Way's scale radius is 7,000 light-years (and corresponding diameter 14,000 light-years).

Galactic disks are in a sense like giant CDs on a player, spinning around fantastically, with their apparent lack of motion due to the enormous distances between them and us. (Although, of course, galaxies are not solid bodies.) Astronomers now know the rotational velocities of M31 as 225 km/s some 1,300 light-years out from the core; a paltry 50 km/s 7,000 light-years from the core; and an impressive 250 km/s some 33,000 light-years from the core, beyond which it slowly declines. Material very close to the galaxy's center, swirling around a central black hole, is moving at about 160 km/s. This velocity is the speed of stars and gas outside the region where the black hole gravity dominates. Close to the black hole, material swirls at speeds that approach the speed of light.

The Andromeda Galaxy's disk is marked by a number of peculiar features that have been uncovered in the past decade. The galaxy appears to contain two broad spiral arms that originate some 1,600 light-years from the hub and are separated in origin by at least 13,000 light-years. Observations from the European Space Agency's Infrared Space Observatory in 1998 suggest the galaxy may be transitioning into a ring galaxy, as large amounts of gas and dust appear in overlapping rings. The Spitzer Space Telescope revealed that M31's primary spiral arms are discontinuous. Spitzer also revealed a small inner dust ring that may have been created by a close interaction with M32 some 200 million years ago.

Like most galaxies, M31 has a condensed star cluster near its center and a supermassive black hole that is quiescent, having run out of a stream of infalling matter long ago. The Chandra X-ray Telescope has studied discrete x-ray sources scattered across the central region, surrounding the black hole at relatively close distances. The original research on deciphering the nucleus of M31 was conducted in 1991 by the American astronomer Tod R. Lauer (1957–). Lauer used the Hubble Space Telescope to find that the nucleus has two concentrations separated by some 5 light-years, and that the dimmer of the two, designated P2, coincided with the black hole. More recently, a 2005 study led by German astronomer Ralf Bender described P1 and P2 as dense collections of stars and a "blue nucleus," designated P3, as the region containing the supermassive black hole, which stands at about 100 to 200 million solar masses. And in 2012 astronomers detected a so-called microquasar in the Andromeda Galaxy – that is, a radio-emitting x-ray binary. These objects produce strong, focused radio emissions with pronounced jets, and contain an accretion disk surrounding a stellar mass black hole.

Outside the galaxy's disk, M31 holds a considerable collection of at least 460 globular star clusters, the brightest dozen or two of which are visible (as starlike sources) with large amateur telescopes. The most massive of these clusters, G1, also designated Mayall II, lies 130,000 light-years from the galaxy's core and is too massive to be considered a normal globular cluster. It holds twice the mass of the bright Milky Way globular known as Omega Centauri (NGC 5139) and may harbor its own small central black hole. This has prompted some astronomers to propose that G1 is the remnant of a captured dwarf galaxy that was stripped of its outer envelope by the gravitational tide of M31.

In contrast to the globular clusters in the Milky Way, which are all very old, M31's globular clusters show a more diverse range of ages. Some globulars show evidence of being only a few hundred million years old, while others are billions of years old, more in line with those of the Milky Way. In 2005, astronomers detected several clusters of stars in M31 containing enormous numbers of stars, hundreds of thousands, in low-density groups spanning several hundred light-years across. These clusters are unlike any known in other galaxies as yet.

The Andromeda Galaxy's overall luminosity has long been a source of interest to astronomers. Not only is M31's disk larger than the Milky Way's, but also it appears to have more low-mass stars. The galaxy's luminosity, at least 26 billion times the solar luminosity, is some 25 percent greater than the Milky Way's overall brightness. In 2010, a study with the Spitzer Space Telescope concluded M31's luminosity might be as high as 36 billion times that of the Sun.

Currently, the Andromeda Galaxy has a relatively low rate of star formation, producing about one solar mass per year, compared with three to five solar masses in the Milky Way. The Milky Way also produces more supernovae than M31, all of which suggests the Andromeda Galaxy was a very active hotbed of star formation long ago but now is in a quiet phase. Plotted on a galaxy color-magnitude diagram, both the Andromeda Galaxy and the Milky Way are now in the so-called green valley, a region between galaxies with active star formation and those that lack star formation. But a future event could certainly spark huge new rounds of star formation – and almost certainly will.

To understand the evolution of knowledge about the Andromeda Galaxy's future, we need to return to the timing argument of Kahn and Woltjers, some 50 years ago. One byproduct of this study was a newfound respect for the amount of dark matter that must surround the galaxies nearby us in space. The dark matter halos around the Local Group galaxies dwarf their bright disks and extend much farther out than even their halos of globular star clusters. But the work of Kahn and Woltjers also ignited a revolution in thinking about the Andromeda Galaxy's future, as well as the past and present of the rest of the Local Group galaxies.

For a long time, astronomers have known that the Andromeda Galaxy and Milky Way are moving closer to each other, as the galaxy's spectrum is blue-shifted. The most recent estimate of M31's radial velocity, its motion toward us, is 300 kilometers per second. Imagine traveling roughly from Chicago to Indianapolis once each second and you can appreciate the velocity very nicely.

Studies of the Local Group membership, and particularly the relationship between the Milky Way and the Andromeda Galaxy, have produced many papers over recent years. One of the most seminal arrived in 2008 and was the work of the American astronomers Abraham Loeb and T. J. Cox at Harvard University. At the time, Cox was a postdoctoral fellow. Realizing that the Milky Way and M31 are heading toward each other now is one thing, but does their velocity mean that motion will continue to carry them together into an eventual collision? The 2008 work by Loeb and Cox set about answering that question. And the answer turns out to be a resounding yes.

Both the Milky Way and the Andromeda Galaxy have larger dark matter halos, and numerical modeling shows such halos exert dynamical friction and act as sponges that "soak up" energy and angular momentum in the dynamics

Figure 10.2 Galaxy mergers like Milkomeda are common in the universe. The Hubble Space Telescope has uncovered a swarm of interacting galaxies in the cluster Abell 1689.
NASA, ESA, the Hubble Heritage Team (STScI/AURA), J. Blakeslee (NRC Herzberg Astrophysics Program, Dominion Astrophysical Observatory), and H. Ford (JHU)

of these galaxies, which would lead to a merger. Previous models broadly hinted at the collision and merger of the two galaxies, which astronomers have dubbed "Milkomeda," but the Loeb–Cox study is the most extensive, and used the most sophisticated modeling using a computer simulation.

Loeb and Cox used a model of the Local Group that assumes normal, baryonic matter is limited to the disks and bulges of the two galaxies, and that a vast dark matter halo around each galaxy has about 20 times the mass of the normal matter. They used total masses for the galaxies of 1 trillion solar masses for the Milky Way and 1.6 trillion solar masses for the Andromeda Galaxy. Unlike the early thoughts about the Local Group's mass, however – that nearly all of it existed between these two giant galaxies – Loeb and Cox argue for a diffuse intragroup medium, a soup of particles (80 percent dark matter and 20 percent gas) that contains another 2.6 trillion solar masses over the whole of the Local Group.

And the fact that this soup of particles exists between the two galaxies means understanding their orbits over time is far more complicated than simply harking back to Kepler's laws. The astronomers settled on a best-fit model that begins by simulating the two galaxies separated by 4.2 million light-years and on an eccentric orbit that brings them to a perigalacticon – a closest separation at initial approach – of 1.5 million light-years.

The astronomers modeled the dynamics between the two galaxies from 5 billion years ago, a short time before the formation of our solar system, to a point 10 billion years in the future, long after the Milky Way–Andromeda Galaxy merger. Not only did they take the overall status of the two galaxies into consideration, but also the fate of our own Sun.

The Loeb–Cox analysis produced some very intriguing results. They found the two galaxies make a first, close passage in less than 2 billion years. The collision and merger itself will take place on a timescale of something less than 5 billion years. The relatively short times involved in the passage and then merger – given the vast future of the universe as a whole – is due to the gravitational effects of the intragroup medium. The merger accelerates and completes pretty quickly once the dark matter halos of the two galaxies penetrate each other, when the galaxies are separated by about 300,000 light-years.

Of course, amateur astronomers can only dream of the night sky that will exist from Earth's viewpoint in the distant future, with the form of the Andromeda Galaxy becoming closer and closer, larger and larger in our sky. It will eventually rival the disk of the Milky Way in size and brightness across our sky, and the resulting show that any living beings will witness will be absolutely incredible. We know from Chapter 3 that by the time these two galaxies merge, Earth will have been inhospitable for a very long time. But perhaps descendants of ours or other intelligent beings throughout both galaxies will bear witness to this transformation.

In 2012, Dutch–American astronomer Roeland van der Marel led a team that used the Hubble Space Telescope to very accurately study the velocity of M31 and added to the study of the coming merger. Their observations spanned a period of 7 years and focused on selected regions of the Andromeda Galaxy, observing the motions of stars. They measured the motions of thousands of stars within the galaxy relative to very distant background galaxies and could thus gauge the stars' "sideways" motion, finding that M31 has virtually no tangential motion and is thus headed straight for us.

The study was a demanding one because it required sophisticated techniques of data analysis to tease the best information from their observations. The angular motions of the stars in M31 over these relatively short periods were tiny and difficult to measure. The astronomers followed up the observations and produced a computer model that in effect eliminated the motions of the stars within the galaxy and looked at M31's overall motion.

The scenario of what is likely to happen in the future would give skywatchers on our planet an incredible view. Two billion years from now, a billion years after the end of life on Earth, M31 will loom larger in our sky, appearing 10 times larger in our sky than it does now. But the gravitational effects between the two galaxies will be minimal at that point.

Three billion years from now, M31 will appear as a very large, hazy glow whose edge intersects the Milky Way in Earth's sky. The gravitational effects will now start to become significant. The compression of gas in the galaxies' disks will trigger a spate of star formation, creating reddish HII regions – areas of new star formation – and star clusters all along the disks of both galaxies.

Four billion years into the future, the Andromeda Galaxy will begin to intermingle with the Milky Way. As this event approaches, M31 will rival the brightness of the Milky Way overall in Earth's sky. Galaxies consist of so much empty space that stars will not collide with each other like cars on a freeway; instead, gas clouds will come together and create a glowing, almost psychedelic pattern in Earth's night sky. Reddish emission nebulae will light up in new fits of star formation, dotting the sky in haphazard patterns. The overall night sky will contain twice as many bright stars as we are accustomed to seeing today.

The stars from the Andromeda Galaxy will move in from a different direction than those of the Milky Way, and the nearest stars will be on high-speed vectors that will carry them as far as the width of the Full Moon in the course of just 10 years. The familiar constellations we now know, most of which will have changed dramatically independent of the M31 merger, will alter over short timescales.

For observers inside the Milky Way, the dramatic transformation that's beginning offers a fascinating look, frame-by-frame, at something we can only imagine. But there is a counter, too. Many galaxies that were previously visible are now blotted out of the sky by intervening gas and dust. Numerous star clusters wink on, supernovae pop like flash bulbs, emission nebulae glow with pinkish hues, and tangled twists of nebulosity and lanes of dust run across the sky in a crazy patchwork.

Some 5 billion years from now, the cores of the Milky Way and Andromeda Galaxy will form a twin stellar bulge in the sky. The definition of exactly what constitutes the center of the galaxy, the now merging galaxies, is unclear. The method used in the past, of determining the galactic center by measuring the distribution of globular star clusters surrounding it, no longer works in a double galaxy governed by chaos.

Following the 5-billion-year mark, the galaxy's clouds of nebulae will begin to diminish in intensity after much of their gas has turned into new clumps of stars. Now distant background galaxies begin to shine through the interstellar medium, which has lost some of its opacity, and we have a chaotic galaxy with a rich background of distant spirals and ellipticals.

Seven billion years into the future, the merger of the Milky Way and Andromeda will be in full throttle. The result will now appear something like the galaxy Centaurus A, a giant, messy elliptical galaxy with chaotic, active features. The crazy flow of materials throughout new and messy orbits will presumably reawaken the central black holes in the nuclei of both galaxies, producing high-energy jets and outflows of x-ray emission as stars and gas are pulled into the accretion disks. The Sun, now a white dwarf, will be a tiny remnant of its former glory.

The name Milkomeda will now be fully earned as both galaxies blend into each other. (As an historical anecdote, Loeb says he devised the name Milkomeda while writing the paper with his colleague Cox. Weary of describing the

Figure 10.3 Forming stars glow fiery bright in this infrared image of the Andromeda Galaxy made with the European Space Agency's Herschel spacecraft. The galaxy's spiral structure will transform into a large elliptical galaxy along with the Milky Way during the Milkomeda merger.
ESA/Herschel/PACS/SPIRE/J. Fritz, U. Gent; X-ray: ESA/XMM Newton/EPIC/W. Pietsch, MPE

product as "the merger product of the Milky Way and Andromeda," he determined to save space with an acronym. Checking Google, he saw the word Milkomeda had not been used. And a name was born.)

Because of the sudden downturn in star formation and supernova explosions, the supergalaxy undergoes a relatively quiet phase in its first ordered eons after the merger is complete. Some of the satellite galaxies from both former spirals, the Magellanic Clouds, M32, NGC 205, and others, survive, having been gravitationally cast out like marbles on a shooting table. Others were absorbed into the maelstrom of the collision. The final stages of the formation of one of the largest galaxies in this part of the local universe are finished.

Eight billion years from now, it is difficult to forecast what the sky will look like from the vantage point of our solar system. Predicting where the Sun will end up, or how it will orbit, in this giant galactic mess is extremely tricky. The Sun currently orbits the center of the Milky Way at a speed of about 220 km/s. The orbital chaos of the Milkomeda collision, however, will increase the Sun's velocity to a number significantly higher than that, which will mean more frequent encounters with other stars passing nearby. That will unleash more comets into the inner solar system, knocked into inward orbits by the stars, as well as by the galactic tide and by passing molecular clouds. Showers of comets will bombard the inner solar system, whether or not anyone is around to see them.

Nine billion years from now, the galaxy will be a fully formed elliptical. Many of the stars within it, perhaps the Sun too, will travel around the galaxy's center on orbits with high inclinations and eccentricities, swooping far out near the galaxy's edge over their long journeys. The view of the combined galaxy will be a magnificent one, although the large, amorphous glow now in the sky will look like a huge, fuzzy searchlight. In Milkomeda's center, the black holes from the Milky Way and M31 will have by now combined into one giant black hole, following a prolonged period of closely orbiting each other. At this time, the black hole has quieted down, again removed from ample gas and stars to pull into its clutches.

The study by Loeb and Cox investigated the possible outcomes for the Sun during the galactic merger. After the first close passage of Andromeda to us some 1.8 billion years from now, the Sun will likely remain in the Milky Way's disk. To be sure, the galaxy's disk will by then be gravitationally disturbed, and the astronomers found a 12 percent chance that the Sun will be ejected from the disk at this stage, finding itself in a tidally ejected wave of material more than 60,000 light-years from the galaxy's center. (The Sun's present distance from the galactic center is 27,000 light-years.)

As the merger process continues, the possibilities of the Sun moving farther out from the galactic center increase. In 3.5 billion years, a second close passage occurs and this will eject the Sun some 30 percent of the time in the Loeb–Cox model. After 4 billion years have passed, the odds of an ejection of the solar system rise to 48 percent and, following the completion of the merger, a greater than 50 percent chance. But the orbits will likely be eccentric and so many stars will spend part of the time far away from the galactic center and part of the time closer to it.

The astronomers found a slight chance that the Milky Way could lose its grip on the white dwarf Sun and our solar system could become gravitationally bound to Andromeda before the merger. This situation would be caused by material becoming loosely bound during the first close encounter and being swept up by Andromeda during the second close passage. But this result is pretty unlikely – Loeb and Cox found it occurred only 3 percent of the time in their simulations.

Clearly, interactions between galaxies that astronomers observe in the cosmos cause waves of new star formation. This is obviously the case with so-called ultraluminous infrared galaxies (ULIRGs), galaxies that emit more than 1 trillion times the luminosity of the Sun in the infrared part of the spectrum. As the galaxies interact, torques take angular momentum from the gas inside the galaxies and push it into the galaxy's centers, initiating a wave of starburst activity.

Figure 10.4 The Hubble Space Telescope's workhorse instrument, the Advanced Camera for Surveys, captured this image of strange interacting galaxies some 300 million light-years away. Appropriately dubbed "The Mice," NGC 4676 appears to be caught in a galactic chase. NASA, H. Ford (JHU), G. Illingworth (UCSC/LO), M.Clampin (STScI), G. Hartig (STScI), the ACS Science Team, and ESA

In fact, Loeb and Cox investigated the possibility of the Milky Way–Andromeda merger producing an ULIRG. Surprisingly, however, they found that the star formation in the merger years will be enhanced only weakly above what it is now. Both galaxies have a relatively low amount of gas, which will prevent a more spectacular starburst event. Further, the astronomers found that most of the gas will already have been consumed by normal star formation by the time of the merger. A potential starburst of a magnificent size will simply lack the fuel to get going.

Interestingly, however, the astronomers found they could not rule out the possibility that the supermassive black holes in each of the galaxies' centers, when merged, might trigger a quasar event. A quasar is the extremely powerful and luminous center of a galaxy marked by high-energy emissions in multiple wavelengths, including x-ray outflow. Loeb and Cox found that while the galaxies lack the gas to fuel a major starburst event, they do have enough gas to fuel a quasar in the center of the newborn galaxy if 1 percent of the gas is accreted by the combined black hole.

The astronomers also forecast the nature of the merger remnant, the elliptical galaxy-like form that will exist when the two galaxies are married. Will Milkomeda dominate a Local Group consisting of a separate M33 and many smaller galaxies, or will the dominant galaxy eventually gravitationally take over and the entire Local Group will be absorbed into one supergalaxy?

Loeb and Cox clearly argued that Milkomeda will resemble a spheroidal, elliptical galaxy and will have stars that orbit in a large range of chaotic ways, and that the galaxy will have a slow rotation. Beyond 10,000 light-years from the center, dark matter will dominate the galaxy. They argue that, depending on the line of sight of a hypothetical viewer, Milkomeda could be seen as "disky" or "boxy," despite the overall spheroidal shape. They also predict Milkomeda will contain a gas-rich halo drawn from the medium surrounding the two galaxies as they come together.

In short, Milkomeda will look like a low to moderate luminosity elliptical galaxy. Because of the low gas density of the two galaxies that created it, however, Milkomeda will have a pretty low density of stars within it. Many elliptical galaxies we see in the present universe have undergone starburst events and so are rich in newly formed stars, which will not be the case in our own future galaxy.

We now know that within 5 billion years the Milky Way and Andromeda Galaxy will merge into a single chaotic galaxy. This is also about the same time frame as the Sun's normal main-sequence lifetime, after which it will become a red giant, a planetary nebula, and finally a white dwarf. Although this will occur long after we believe Earth can sustain life, it is probable that many other sentient beings, perhaps even our descendants, will witness this great collision of galaxies.

If they are around, future astronomers will need to observe the merger with great enthusiasm. One day, some 100 billion years from now, the accelerated expansion of the universe will pull all visible galaxies except those in the Local Group too far away to be visible at all. If the cosmological constant does not evolve over time, then 100 billion years from now extragalactic observational astronomy will come to a gradual halt. What remains of Milkomeda will then be the only part of the universe anyone in this region of space could possibly see. Thankfully, there will still be plenty to observe in Milkomeda for far-future astronomy enthusiasts.

Chapter 11
The Big Bang's cosmic echo

Few things have polarized the astronomy enthusiast community like the Big Bang Theory. Despite overwhelming evidence for the Big Bang origin of the cosmos, a select group of amateur astronomers resists accepting it, although this strange phenomenon has dimmed somewhat as more and more evidence for the Big Bang has piled up.

The basis for the logic behind the Big Bang is simple. In the 1910s and 1920s, Vesto M. Slipher, Edwin Hubble, and others discovered initial evidence for the expansion of the universe – everything appears to be expanding away from everything else on large scales. Running the history of the universe backwards, then, naturally brings all of this expanding material together into a very small initial space.

The initial, primeval, incredibly dense state, in which all matter and energy were compressed almost unbelievably and then expanded as space-time rushed outward, is now familiar as the Big Bang. But such was not always the case. In 1922, Russian physicist Alexander Friedmann (1888–1925) led the charge with an idea that perhaps the universe had an initial singular state of extraordinarily high density. Five years later the Belgian priest–astronomer Georges Lemaître (1894–1966) published a milestone paper suggesting a dense origin of the cosmos and a "primeval atom." Lemaître believed the dense early universe was akin to a huge radioactive atomic nucleus. Because Lemaître was a priest, some cosmologists were initially skeptical of his notions.

In the late 1940s, the adventurous English cosmologist Fred Hoyle (1915–2001) proposed a major alternative to what was shaping up as the Big Bang. Calling it the steady-state theory, Hoyle and his colleagues, Thomas Gold (1920–2004) and Hermann Bondi (1919–2005), suggested that matter is created continuously as the universe expands, allowing the cosmos to be homogeneous (alike everywhere) and isotropic (uniform in all orientations) in space and time. Hoyle, in fact, appeared on BBC Radio in 1949, arguing against the Big Bang Theory, and using the term "big bang" for the first time, derisively (or perhaps jokingly). To his utter horror, the term stuck and was used by the Big Bang's proponents.

The uncertainty over differing cosmological models lingered for a variety of reasons. Working against the Big Bang was the fact that the Hubble Time, an indicating figure of the age of the universe, seemed to be less than the age of Earth. That problem ebbed away as further observations of distant galaxies in the 1950s by German astronomer Walter Baade (1893–1960) and American astronomer Allan Sandage refined results.

In the late 1940s, another group of researchers began a consistent program of studying Big Bang cosmology. It was led by Russian–American astrophysicist George Gamow (1904–1968), also including his American colleagues Ralph Alpher (1921–2007) and Robert Herman (1914–1997). The group led by Gamow argued in 1948 that the early universe was not only incredibly dense but also incredibly hot, and this led to understanding that an era in the early universe was dominated by intense radiation, not matter. They also realized that this era of radiation lived on; the radiation cooled over time and spread throughout space-time, and this radiation should be detectable as an afterglow of the Big Bang. They named the radiation the cosmic microwave background (CMB), and estimated its present-day temperature at 5 to 50 kelvins.

For many years, proponents of the different cosmological models battled with each other, trying to win over influence for their ideas of the early universe. The argument effectively ended in 1964, when two radio astronomers working at Bell Laboratories in New Jersey accidentally discovered the cosmic microwave background radiation, the snapshot of the oldest light in the cosmos, imprinted onto the sky, with their observations in the microwave portion of the radio spectrum. Arno Penzias (1933–) and Robert Wilson (1936–) won the Nobel Prize in Physics in 1978 for their discovery, and a litany of other discoveries and spacecraft data since have added details and further cemented the conclusion that the Big Bang model explains the early history of the universe.

The best data we have on the age of the universe, from the European Planck satellite and announced in 2013, reveal a cosmic age of 13.798 ± 0.037 billion years, so let's call it 13.8 billion years. That is a pretty high precision number. And the Planck satellite also has revealed much about the composition of the universe – assisted by other studies – that 4.9 percent of the matter-energy in the cosmos consists of normal, baryonic matter (all the bright stuff we can see – stars, galaxies, planets, trees, cats, us, and so on), 26.8 percent consists of dark matter, and 68.3 percent consists of dark energy. More on all of this fun stuff later.

For now, we will take a journey backward in time to explore elements of what has happened in the history of the cosmos according to Big Bang cosmology, for which there is ample evidence. We know that the average matter density is inversely proportional to the cube of the size of the universe. So when the cosmos was one-tenth its present size, the density was 1,000 times greater. In the present-day universe, the average density is roughly 1 hydrogen atom per cubic meter, but in the earliest days of the universe, the density was a million or more times greater.

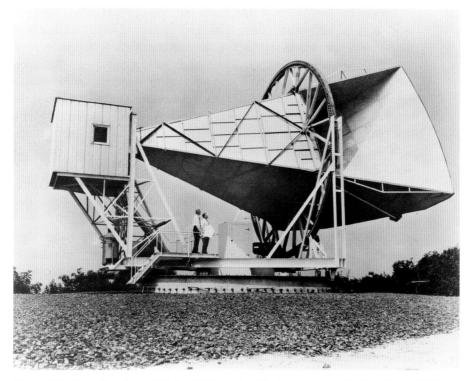

Figure 11.1 Arno Penzias and Robert Wilson stand in the foreground of the Holmdel Horn, owned by Bell Labs, which they used to serendipitously discover the cosmic microwave background radiation in 1964, a find that would eventually win them the Nobel Prize. NASA

The history of the cosmos traveling back in time creates a fascinating story. Some 4.6 billion years ago, we see in our well-formed galaxy, the Milky Way, a young open star cluster that contains a G2 star, the Sun. Twice as far back in time, we can see the Milky Way itself forming as its stars, gas, and dust come together, rearranged by the mergers of many smaller galaxies that came before it. Much farther back in time, when the universe was only a billion years old, we come to the era when gas and dust were assembling by gravity to form the first large aggregates of galaxies. The Hubble Ultra Deep Field images show us a great glimpse of this cosmic era, when small galaxies were merging to form larger ones.

Earlier still, back to the cosmic age of just 1 billion to 400 million years, we see the first stars and galaxies forming in the so-called era of reionization, when a major phase transition occurred, ionizing hydrogen gas and allowing lots of ordinary matter to assemble into huge structures of galaxies and stars. In this day, assemblages of gas come together by gravity to form the infant beginnings of what will grow into magnificent elliptical and spiral galaxies later in the universe.

Before this time, there exists the mysterious period called the cosmic dark ages, a time before the first stars and galaxies, when darkness prevailed throughout space-time. In this era, some 800 million to 150 million years after the Big Bang, most photons in the cosmos were interacting with electrons and protons in a "fluid" of photons and baryons, making the universe opaque. The cosmos was subsequently cloaked by fog, and no light was emitted during this era that we can now detect. This era may have been dominated by primordial black holes.

At an age of 15 million years, the universe emitted a room-temperature glow of 300 kelvins (27 °C, 81 °F) – bathed everywhere in the radiation of a warm summer afternoon. When it was only 1.5 million years old, the universe was bathed in a reddish glow and was rather warmer than a sunny afternoon, at 1200 kelvins (927 °C, 1700 °F).

At an age of approximately 380,000 years after the Big Bang, the universe reaches a crucial moment in his history. This age of recombination, as it's called, marks a turning point. Before this stage, atoms of hydrogen and helium were ionized – that is, consisting of positively charged nuclei but without electrons. As the universe cooled to a critical state, these electrically charged nuclei captured electrons, in the so-called era of recombination. This changed the universe such that most protons were bound in neutral atoms, freeing photons in near-infinite paths, in a process called decoupling. This created the imprint of light called the cosmic microwave background radiation, the same light detected by Penzias and Wilson and studied by a battery of instruments and spacecraft ever since. Hence, the cosmic microwave background radiation gives us a picture of the universe at the end of this important era.

Moving even farther back in time takes us into the province of the early universe, where the chronology becomes even stranger. In the oldest moments of the early universe, we find a state of matter in which atoms don't exist – the temperatures are too high for atoms to hold together, and in this so-called ionized era, atomic nuclei exist in a huge bath with decoupled electrons. The universe is a giant cocktail sea of plasma, and this state of freely mixed particles sustains for about 380,000 years. The fact that the atoms were ionized unleashes a spray of electrons everywhere, meaning photons cannot travel freely and the cosmos is locked in an opaque glow. This prevents astronomers from seeing back any farther in time than the era of decoupling. The light they do see from this era is the cosmic background radiation, the imprint of the early universe. Before the era of decoupling, radiation was scattered continuously by free electrons. At the decoupling era, the universe glowed uniformly at 3,000 kelvins, hot enough to melt titanium.

Another crucial aspect of the early universe was the changeover of relative mass densities of matter and radiation. In the so-called era of equal densities, every cubic centimeter of space contained the same mass as equivalent energy (using $E = mc^2$). As astronomers look back into the universe, of course, they see the density of matter increasing. But the density of radiation increases even

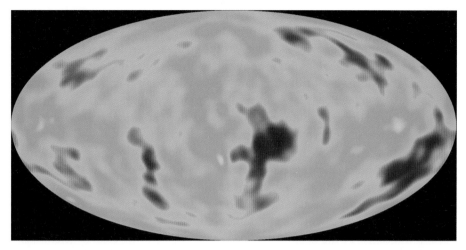

Figure 11.2a The cosmic microwave background shows the beginnings of structure in the universe, with color differences indicating density fluctuations of only a few parts in 100,000.

Figure 11.2b As the universe evolved, these would become galaxy clusters and voids. In 2013, the European Space Agency released a CMB map from their Planck spacecraft that significantly improved upon the work NASA had started 20 years earlier with the Cosmic Background Explorer.
NASA (COBE); ESA and the Planck Collaboration (Planck)

more quickly than that of matter as we look back in time. At around 100,000 years after the Big Bang, the radiation and matter densities were equal. Along with the decoupling era, the era of equal densities occurred near the end of the early universe.

The bulk of the early universe consisted of the radiation era, when the universe was dominated by radiation. This crucial era started very quickly after the Big Bang and lasted for approximately 300,000 years. At the end of the radiation era, when matter and radiation had similar densities, they each had the equivalent of 3 thousand hydrogen atoms per cubic meter – about 3 thousand times the present density of the cosmos. Farther back in time, the density of radiation springs high above that of matter. This now brings on the latter stages of the era of the Big Bang itself. Although this universe is dominated by radiation, which glows incredibly brightly, matter plays an important role because uncountable free electrons scatter radiation everywhere, causing the universe to behave like a giant fluid.

In the first few hundred seconds of the radiation era – following that first second after the Big Bang – about 25 percent of all matter transforms into helium nuclei. In the more than 13 billion years that have followed this conversion, the universe has labored to transform hydrogen and helium into heavier elements. And yet over this incredible span of time, only 2 percent of all hydrogen in the cosmos has been transformed into heavier elements. Now set aside this incredibly long transformation of hydrogen – in the first few hundred seconds of the radiation era, the universe converted more than 10 times the hydrogen into helium than it has since. In the early part of the radiation era, the energy of the cosmos is tied up in what is in effect a gigantic hydrogen bomb – but the conversion was independent of the universe's expansion.

On a particle scale, for a short time, the universe plays a never-ending game of ping-pong. It continuously assembles neutrons and protons, forming

Figure 11.3 Two instruments housed at the South Pole's Dark Sector Laboratory, BICEP2 and the South Pole Telescope, are seeking similar signs of echoes leftover from the Big Bang. Steffen Richter/Harvard University

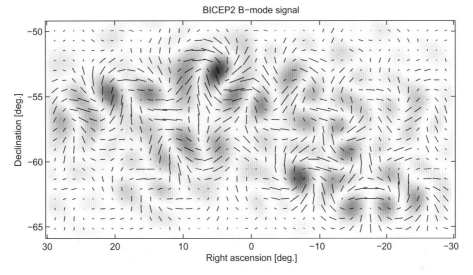

Figure 11.4 Scientists working on the South Pole's BICEP2 instrument imaged a twisting pattern on the sky, which they interpreted as a sign of cosmic inflation in the universe's first second. Red and blue shading indicates the degree of clockwise and counter-clockwise twisting of this so-called B-mode pattern.
BICEP2 Collaboration

deuterons, the nuclei of so-called heavy hydrogen, deuterium. But just as quickly and continuously, the deuterons are stripped apart, splaying the particles into individual neutrons and protons once again. This was the case in the first seconds. After 100 seconds, the universe's temperature drops to 1 billion kelvins and the radiation can no longer break apart the deuterons. This enables the deuterons to form helium nuclei in a burst of energetic formation that lasts 200 seconds, creating the 25 percent of matter now consisting of helium. Except for a small amount of deuterium, helium-3, and lithium, the rest of the cosmos is still hydrogen.

Still farther back in time, we find the so-called lepton era, from 1 second to 10 seconds after the Big Bang. Leptons are very light particles, such as electrons, positrons, and neutrinos. When the lepton era begins, the universe is aglow at a temperature of 1 trillion kelvins, and the density reaches 1 billion grams per cubic centimeter. At the end of the lepton era, the beginning of the radiation era, the temperature has cooled to 1 billion kelvins and the density to 1 million grams per cubic centimeter.

During the lepton era the universe is feverishly busy producing electron pairs, consisting of electrons and positrons, and neutrino pairs. Energy and interaction is ceaseless here, with pairs of electrons and positions constantly being created and annihilated in a nonstop game of particle life and death. Nucleons (protons and neutrons, the builders of normal matter) are also here, but buried deep within a sea of particle froth. For each nucleon at this stage,

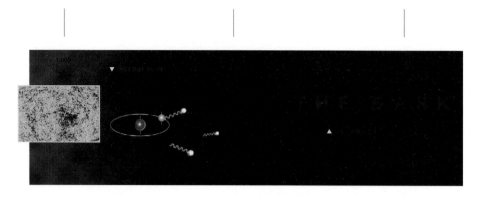

Figure 11.5 The cosmic microwave background formed as the universe cooled enough for electrons to join with protons. This let CMB photons travel through the universe unimpeded. *Astronomy*: Roen Kelly

there exist a billion photons, electron pairs, and neutrino pairs. As far as we know, protons are stable – they can exist essentially forever, and so there are more protons than neutrons today.

There are other exotic kinds of leptons, too – so-called muons, which are similar to electrons but heavier, and like electrons can also possess their own neutrinos. Muons decay in a flash into an electron and neutrinos. Traveling back further in time brings into play a vastly higher temperature of about 10 billion kelvins, where enormous amounts of electrons and neutrinos are produced; only when the temperature is 1 trillion kelvins can muons also enter the scene.

At the start of the lepton era, when the cosmos is 1 second old, the universe is awash in photons, electrons, muons, and the neutrinos from electrons and muons, as well as antimuons and tau neutrinos. As the universe expands and cools, the muons eventually collide and annihilate each other, exiting the stage. Also as the universe cools, to about 5 billion kelvins, enormous numbers of electrons begin to vanish, their energy taken up by photons.

Neutrinos play a key role in the lepton era, both at its beginning and end. They are weakly interacting particles that have very small mass and travel near the speed of light, and have no electrical charge. At the beginning of the lepton era, muon neutrinos decoupled, and at the end of the era, electron neutrinos decoupled. Neutrinos from the early universe are still around, travel at light speed, and form a cosmic background glow of their own. Amazingly, each cubic centimeter of space holds on average 400 background photons and 300 background neutrinos. But neutrinos do not like to interact with matter. Detecting these primordial neutrinos will give us a window into the first second of the cosmos.

Going back even further, to the first microseconds of the cosmos, reveals a universe of ever-stranger character. At 100 microseconds, the universe has a temperature of 1 trillion kelvins and its density is 100 trillion grams per cubic centimeter – the radiation density, the equivalent of 100 million tons of Earth

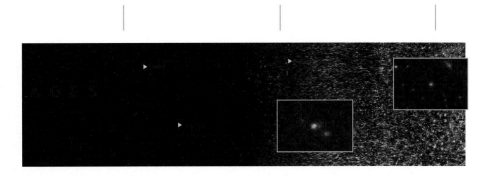

Figure 11.5 (*cont.*)

rock crammed into a thimble. This is agonizingly close to the Big Bang itself, but much of cosmic history is encoded in the first fractions of a second. In this sea of particles so close to the Big Bang lies the hadron era, which includes baryons of matter (protons and neutrons) and field particles, or mesons, which include pions and kaons. In the era of hadrons – which means strongly interacting particles, referring to protons and neutrons – the universe is bathed in a sea of hadrons because the high temperatures enable their creation. This soup of particles contains an immensely dense ocean of photons, leptons, and hadrons – all with their antiparticles – annihilating each other and constantly being recreated.

But the story is fluid – no pun intended. As the universe cools, the annihilation of particles wins out and toward the end of the era matter is almost wiped out universally. Most all antimatter, the antiparticles, is gone. But a slight imbalance, with more matter than antimatter, means the normal particles won out – were not all annihilated – and the universe soldiered on. The difference was incredibly slight, however, with about one part per billion more matter than antimatter. Amazingly, one of the incredible facts we learn from the very early universe is that the number of particles that existed then and that exist now is roughly the same. Increasing entropy, that disappointing concept we all learn about early on in school – that the cosmos goes from order to disorder over time, that matter is breaking down and becoming disorganized, is a fact – and the nature of the particles, the matter, changes over time. But the numbers of particles are roughly constant over time.

What this means is that the universe, over its long history, has transformed dramatically. Early on, the young universe consisted of nearly equal parts of matter and antimatter, with that slight imbalance toward matter of an extra part per billion. That small difference, however, is what led to us – all the baryonic matter we know of, including stars, galaxies, planets, and human beings. Without the imbalance, the cosmos would only have consisted of photons and neutrinos, and we would not exist.

Continuing to travel backwards in time, on toward the Big Bang itself, is very difficult. We understand the nature of things down to about 10^{-12} second, but beyond that, difficulties arise. Physicists enter a world in which the familiar laws we understand, those created by Newton, Einstein, and their followers, become progressively more difficult to maintain as governors of the situation.

For the later universe, we can understand the laws of physics by focusing on the four fundamental forces – the gravitational force, the weak force, the strong force, and the electromagnetic force. Since the era of the hadrons, these four forces have governed actions in the cosmos.

But the earliest period of the cosmos following the Big Bang, the so-called Planck Era, offers mysteries. This period extended from the Big Bang itself until approximately 10^{-43} seconds, known as Planck time. During this brief period, some 13.79 billion years ago, quantum gravity probably dominated physical interactions in the cosmos, and gravitation was as strong as the other fundamental forces. Rather than four discrete forces, physicists believe a single superforce dominated the universe. At the end of this era, the superforce probably split into two forces, gravity and a hyperweak force. And the hyperweak force can change matter into antimatter.

And the understanding of forces in the very early universe becomes even stickier. According to so-called Grand Unified Theories, the universe soon cooled to 10^{28} kelvins and then the hyperweak force split into two other forces, the strong force and the electroweak force. This transition, then, possibly marked the onset of cosmic inflation, the period that lasted from about 10^{-36} to 10^{-34} seconds, during which the universe hyperinflated, which helps to explain a variety of observed cosmic phenomena we see much later on in the history of the universe. Inflation theory was proposed in the 1980s by American physicist Alan Guth (1947–) and Russian–American physicist Andrei Linde (1948–), and others, and enjoys very widespread support by cosmologists. In 2014, hard evidence for cosmic inflation was presented by a team operating the BICEP2 telescope cosmic microwave background experiment at the South Pole, detecting B-mode polarization via gravitational waves, but after a period of several weeks, the scientists backed away from their certainty about the claim.

After inflation, the universe was ruled by gravity, the electroweak force, and the strong force. Finally, at 10^{-10} seconds, the electroweak force split into the weak and electromagnetic forces, and we carried on with the four forces that are familiar to physics today.

An important aspect of understanding hadrons comes from quark theory, which was proposed independently by American physicist Murray Gell-Mann (1929–) and Russian–American physicist George Zweig (1937–) in 1964. Quark theory simplifies the incredible complexity of hadron theory, by suggesting that hadrons consist of basic particles called quarks. Hadrons, whether baryons or mesons, goes the theory, consist of as many as six "flavors" of quarks, which parallel the number of leptons. And each of the six flavors can have three distinct colors, making 18 possible quarks, the flavors being up, down, charm, strange, top, and bottom.

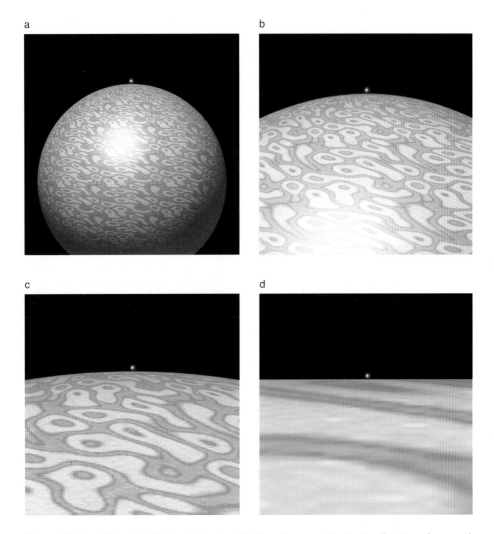

Figures 11.6a–d Cosmic inflation is theorized to have happened just a tiny fraction of a second after the Big Bang. In each panel here, a curved surface is shown expanding by a factor of three, but by the last panel, the view of the sphere looks flat. Astronomers use inflation to explain why the universe is geometrically flat, and the cosmic microwave background is almost the same temperature everywhere.
Astronomy: Roen Kelly

In the very early cosmos, the universe was awash with quarks and leptons. Scientists have not yet isolated quarks individually, but they have detected evidence of their existence. Isolating a quark may be an impossible task, akin to trying to separate just one pole of a magnet. But we know quarks played a key role in the very early universe. Before the hadron era, the quark era sees a cosmic ocean of overlapping particles that cannot be separated, quarks bubbling together in a dense fluid of overlapping matter.

Before the quark era, when the universe was younger than 10^{-36} seconds, it is very difficult to get a clear picture of what the universe was like, because theoretical uncertainties abound. We believe that the Big Bang consisted of all energy and matter in a singularity, a nearly infinitely dense point of origin for the universe we know and that we live in. But understanding those earliest moments of the cosmos poses a hellish challenge – when we go back in time to 10^{-43} seconds after the Big Bang, to the Planck Time, we hit a wall, sometimes called the Planck Barrier.

Now we are in the world of quantum cosmology. At this moment, 100 million trillion trillion trillionth of a second after the Big Bang, the density of the universe is 10^{94} grams per cubic centimeter and the temperature 10^{32} kelvins. Physicists, cosmologists, and astronomers are stopped dead in their tracks at this moment in time, as they know almost nothing about the nature of the cosmos at this time, not to mention before it.

Space, time, energy, and matter are all inextricably tangled at this moment, and the structure of the universe then is completely unknown – the great American cosmologist John Archibald Wheeler (1911–2008) described the cosmos at this era as a giant cosmic foam. Space-time may then have existed as a network of countless black holes. Space and time themselves did not exist in the senses that we now think of them. The entire universe is one big medium – "here" and "now" become meaningless concepts in the very early universe. In the era of quantum cosmology, history no longer exists in the conventional sense, and discoveries that we have yet to make will sort out, we hope, the nature of the universe in the earliest fractions of a second following creation. It should be noted that many cosmologists consider the Big Bang to be inflation, at 10^{-35} seconds. Prior to that, there was just a small packet of energy, not necessarily expanding quickly. But others think of the Big Bang at time = 0.

For now, we need to be satisfied with knowing that the Big Bang happened. We also understand the history of most of the universe right back to nearly the time of the Big Bang. We need to leave that fraction of a second at the Planck Barrier and earlier as one of the great, staggering cosmic mysteries that lingers on.

Isn't cosmology fun?

Chapter 12
How large is the universe?

It is absolutely amazing to know that shortly after the Big Bang, the universe was a relatively small, nearly infinitely dense place. It boggles the mind. But that was 13.8 billion years ago. The expanding universe means the entirety of what we know is now incredibly large, and is getting larger every day.

This is one area that two generations of science fiction movies have seriously distorted in the minds of the public at large. The general feeling that technology is pretty good and will know almost no bounds, and that we can almost certainly one day travel between star systems, are pretty much taken on faith. But what the sci-fi movies have failed to communicate, among other things, is that the universe is an INCREDIBLY LARGE place. Even distances between the very nearest objects are staggering, and the distances across the Milky Way Galaxy and certainly between galaxies in the universe are astonishingly huge to living beings stuck on a planet. A model of the Milky Way wherein the Sun is a grain of sand brings this home. On this scale, stars – sand grains – are 6 kilometers apart in the Milky Way's disk and the disk is about 60,000 kilometers across. Now who wants to go traveling from grain to grain?

The concept of the size of the universe has taken a huge stride forward in just the last few years. There was a time not too long ago when astronomers did not know even the approximate size of the cosmos with any degree of accuracy. We still don't know with high precision. The universe may be infinite.

The Big Bang Theory reminds us that once the universe was very small. We know the fastest radiation or any information can travel is the speed of light, 300 million m/s (186,000 miles per second). We also know the universe is 13.8 billion years old. We also know that a light-year is equal to approximately 9.4 trillion kilometers, or 6 trillion miles. In nearly 14 billion years, on first blush, we might expect radiation to expand radially outward to something like 30 billion light-years across.

Remember that the Big Bang was not like an explosion that went off in a room. Following the Big Bang, space-time itself expanded radially outward

Figure 12.1 The entirety of Earth is contained by just a few stray bright pixels in this rare scene captured from a distance of 898 million miles by NASA's Cassini mission, which has explored Saturn for more than a decade.
NASA/JPL-Caltech/Space Science Institute

at all points, meaning all of space expanded too, not just the stuff within it. (The term space-time refers to the mathematical model that combines space and time into a single, interwoven medium.)

As the expansion of the universe began, that 1 centimeter of "empty space" interstitially became 2 centimeters over time, and so on. So the best ideas about the size of the universe allowing for its expansion over time point to a radius of slightly more than 46 billion light-years, and therefore a diameter for the universe of approximately 93 billion light-years.

And there is a major proviso to that result. That refers to the visible universe we can see from Earth. Inflation theory, if correct – and recall the very wide-spread, overwhelming support for it among cosmologists – suggests the portion of the universe we can see is by no means the entire cosmos. Some cosmologists suggest the universe is in fact infinite. But let's work with what we really have and say the universe, at least the part that we can observe, is about 93 billion light-years across.

We have discussed, very briefly, the distance scale of the objects close to us in space. But a thorough understanding of our neighborhood, our solar system, our area of the Milky Way, our galaxy, and so on, is critical to comprehending how the universe works. And exploring the cosmic distance scale also unveils a slew of interesting objects that astronomers use to determine distances to objects near and far.

The seeds of measuring the universe stretch back in time all the way to the Greek astronomer Aristarchus of Samos (ca. 310 BC–ca. 230 BC), who had correct notions of parallax in mind with regard to distances of the Sun and Moon. Parallax is the technique of measuring the offset of nearer bodies to the distant background of stars, and geometrically calculating a distance. Little progress took place until after Polish astronomer Nicolaus Copernicus (1473–1543) proposed the heliocentric model of the cosmos, and it was one of the last great visual astronomers, Danish nobleman Tycho Brahe (1546–1601), who made the first parallax measurements of comets and helped to define a more modern distance scale to nearby objects.

Now, of course, we know the distances to the Sun, Moon, and planets very well, in the case of the Moon to within inches, through a battery of sophisticated techniques, including laser ranging and an enormously well-understood heritage of orbital dynamics that began with Tycho's assistant, Johannes Kepler (1571–1630) producing his laws of planetary motion.

Earth's Moon, representing the most distant body to which human beings have traveled, is the closest significant object to us in space, lying on average 380,000 kilometers (236,000 miles) from us, meaning that light takes 1.3 seconds to travel from the Moon to Earth.

From the Sun, our star and the central point of the solar system, the planets with their moons and asteroids splay out in a giant wheel, arranged in a beautiful orbiting disk with assorted oddball interlopers and objects such as comets on highly inclined, eccentric orbits that have been spun into a flurry. The innermost planet, Mercury, is some 50 million kilometers (30 million miles) from the Sun. Moving outward, Venus is 108 million kilometers (67 million miles) away, while Earth is 150 million kilometers (93 million miles) from our star. It takes light 8.3 minutes to reach us from the Sun.

Moving outward, Mars lies some 220 million kilometers (136 million miles) from the Sun, and the main-belt asteroids lie in the gap between Mars and Jupiter. The largest planet in the solar system is on average 770 million kilometers (480 million miles) from the Sun. Next comes the beautiful ringed planet Saturn, which lies about 1.45 billion kilometers (900 million miles) from the Sun. The ice giants Uranus (2.85 billion kilometers; 1.8 billion miles) and Neptune (4.5 billion kilometers; 2.8 billion miles) come next. Then a slew of smaller, icy objects usher in the Kuiper Belt and the Scattered Disk. They include Pluto, whether a planet or dwarf planet (as distant as 7.3 billion kilometers, 4.5 billion miles), and other similar bodies, including Eris, Makemake, and Haumea.

Planetary scientists like to use astronomical units as measurements for larger distances in the solar system – 1 AU is the Earth–Sun distance of 150 million kilometers. The planets are arrayed in a line of impressive distances: Mars at about 1.5 AU, Jupiter at about 5 AU, Saturn at about 9.5 AU, Uranus at about 19 AU, Neptune at about 30 AU. The Kuiper Belt of icy objects lies between 30 and 50 AU, and the Scattered Disk, energized comets and icy asteroids that have been spun up into weird orbits by encounters with Neptune, extends to 120 AU. But that extraordinary distance, more than 100 times the distance between Earth and Sun, hardly reaches toward the edge of even our own solar system.

The most remote products of humanity are also in the distant solar system equation – the Voyager 1 and Voyager 2 spacecraft. Launched in 1977, the two Voyagers conducted a "grand tour" of solar system exploration that included visits to Jupiter, Saturn, Uranus, and Neptune, and their primary missions came to an end in 1989. They have been traveling outward ever since, though, at speeds of about 15 km/s, relative to the Sun. In 2013, NASA scientists announced that Voyager 1 crossed the heliopause, the boundary where the solar wind is stopped by the interstellar medium, and thus crossed into "interstellar space." The Voyagers are now at distances of about 128 AU from the Sun (Voyager 1) and 105 AU (Voyager 2). However, there is a game of semantics going on here; while it's true that Voyager 1 crossed the heliopause, the outer gravitational boundary of our solar system is much farther out.

We need to travel a long, long way to encounter the Oort Cloud, the huge sphere of possibly 2 trillion comets that surrounds our solar system and defines the outer boundary of it. Attached to the Sun by gravity, and traveling along with us in our ride about the galactic center, this shell is the cause of occasional comets falling inward toward the inner solar system, where they give us a brief and exciting show in our skies. The inner edge of the Oort Cloud lies some 10,000 AU from the Sun, and the cloud's outer edge some 10 times farther away yet, 100,000 AU, or about 40 percent of the distance to the nearest star system beyond the Sun, Alpha and Proxima Centauri. That enormous distance means that in a gravitational sense, Voyager 1 has traveled only one eight-hundredth of the way to the physical edge of the solar system, and so it won't pass into interstellar space beyond the Oort Cloud for another 28,000 years.

Let's pause for a moment to appreciate the physical scale of just our solar system – only the Sun, its attendant planets and debris, and our little island of life inside it. To envision our immediate vicinity a little better in your mind, imagine a scaled solar system with the Sun on one end and 1 centimeter representing the distance between our star and Earth. That is, 1 AU = 1 cm. You can actually draw this out on paper to help crystallize it in your mind. Tape several sheets of paper together and have a go at it. With the Sun at one end, Earth is 1 centimeter away, and Mercury and Venus are there too at 0.4 centimeters and 0.7 centimeters. Outward from Earth, we have Mars at 1.5 centimeters, the main-belt asteroids centered around 2.5 centimeters, Jupiter at 5 centimeters, Saturn at 9.5 centimeters, Uranus at 19 centimeters, and Neptune at 30 centimeters. (If you read the Pluto chapter, you can put it in at 40 centimeters.)

The outer solar system is sparse, consisting of the Kuiper Belt region from 30 to 50 centimeters from the Sun, and you can even indicate some of the more

Figure 12.2 Earth is a runt in comparison to its strange planetary siblings in the outer solar system, whose orbits stretch out more than 40 astronomical units – 40 times the distance between Earth and the Sun.
NASA

interesting objects in the area to keep Pluto company – Haumea at 40 centimeters, Makemake at 45 centimeters, and Eris at 60 centimeters. Now you can finish by indicating the region of the Scattered Disk between 50 and 100 centimeters from the Sun. This gives you a complete scale model of the solar system in a region spanning 1 meter, or three feet, across. Now appreciate that on this scale, the inner edge of the Oort Cloud is 100 meters (109 yards, more than an American football field) farther away than the edge of your diagram. The outer edge of the Oort Cloud, on this scale, is 1,000 meters (six-tenths of a mile, more than 10 football fields) away. And as human astronaut-explorers, we have traveled as far away as the Moon, about 1/389th of an AU, or on our scale 1/389th of a centimeter, from Earth, which on this scale is about the size of a human red blood cell. That distance is imperceptibly close to Earth's "dot" on our scale drawing.

And yet the distances to the nearest stars are vastly, unbelievably, larger than our imagined scale of the Oort Cloud. We now need to drop astronomical units, the Earth–Sun distance, and think in terms of light-years, the distance traveled by light in a year, about 9.5 trillion kilometers (6 trillion miles), as photons stream through the cosmos at 300,000 kilometers per second (186,000 miles per second). (For those who need to know, 1 light-year = 63,241 AU.)

As we have seen, the nearest star system beyond the Sun is composed of a triple system, the bright Southern Hemisphere signpost Alpha Centauri. This star lies 4.3 light-years away – some two and a half times farther away than the outer edge of the Oort Cloud – and is composed of a sunlike star with a smaller companion orbiting it, and a third, distant star gravitationally bound as well. This third star is Proxima Centauri, which actually lies a little closer to us, at 4.2 light-years, and is a reddish main sequence star. The fastest spacecraft we have sent out into the cosmos, Voyager 1, would take the better part of 70,000 years to travel an equivalent distance from the Sun to Alpha Centauri.

Other nearby stars in our immediate neighborhood of the Milky Way give us an intriguing sample of stellar types. They include Barnard's Star, a faint reddish star 6.0 light-years away that speeds across our sky with a high proper motion, moving relatively quickly compared with the distant background stars. Next comes Wolf 359 in Leo, another faint reddish star. Sirius, the brightest star in our sky, is bright largely because it is so close, at 8.6 light-years. Other close stars include Procyon (the brightest star in Canis Minor), Epsilon Eridani, 61 Cygni, and Epsilon Indi, mostly stars that are much less powerful than our Sun.

Beyond a dozen light-years, many more stars and other inhabitants of our Milky Way Galaxy begin to show up. Recall that our Milky Way has a bright disk component that stretches about 100,000 light-years across, and that our Sun and solar system lie about 27,000 light-years out from the galactic center. The galaxy's thin disk is approximately 3,000 light-years thick. The galaxy's central bar extends some 10,000 light-years long. From its ends originate the long and winding spiral arms that contain many regions of stars, gas, and dust. The galactic center hides Sagittarius A*, the quiescent black hole that lies at the heart of a dense region of matter. The galaxy's halo extends out some 200,000 light-years on either side of the galactic center.

As we explored in the Milky Way chapter, the distances to many objects within our galaxy are not necessarily known to high precision. From our vantage point in the Orion Spur, however, we can gain a good view of millions of objects within our galaxy and of millions of galaxies outside the Milky Way.

When we look toward the galactic center, we see the constellations Sagittarius and Scorpius in our sky. Looking in the opposite direction shows us the bright stars of Orion, and in between, on either side of a full rotation, lie Cygnus and Cepheus on one side and Vela and Centaurus on the other. The galaxy is richly studded with deep-sky objects – interesting stars, star clusters, and nebulae. As we have seen, however, when looking toward the center of the galaxy in visible light, we see only a fraction of the way toward the black hole, only a few thousand light-years away, as most of the space between us and the center is heavily obscured by dust.

Nonetheless, we can clarify the distances to some of the most interesting objects lying around us in the galaxy, which helps us appreciate the distance

Figure 12.3 Recent measurements have pinned down the distance to the Pleiades, a group of stars long watched by humanity. Light from stars in the cluster would have departed for Earth some 420 years ago, about a decade before Galileo first trained his telescope on them.
NASA, ESA and AURA/Caltech

scale. Star clusters are often relatively easy to measure distances to, as they offer equal-aged groups of stars that can be analyzed. The closest star cluster to us, the Hyades in Taurus, lies some 150 light-years away. It is recognizable as the bright, V-shaped grouping of stars that makes up much of Taurus' form, along with the bright orange star Aldebaran (which is closer to Earth). The Pleiades cluster (M45), close to the Hyades in our sky, is about 420 light-years away. Another close open cluster is the Coma Star Cluster (Melotte 111), which forms a hazy core of Coma Berenices, and lies 280 light-years away.

Of the approximately 1,100 open clusters known in the Milky Way, many lie at progressively larger distances. The Beehive Cluster (M44) in Cancer is some 580 light-years distant. The Double Cluster in Perseus (NGC 869 and NGC 884) consists of two groups, similar in appearance, each lying about 7,500 light-years from Earth. M41, a bright open cluster in Canis Major, lies some 2,300 light-years away. The big, bright open clusters M6 and M7 in Scorpius lie at distances of 1,600 light-years and 1,000 light-years, respectively. The Wild Duck Cluster (M11) in Scutum is some 6,200 light-years away. And the huge, sparse cluster M39 in Cygnus is a mere 800 light-years distant. In the Southern Hemisphere sky, no star cluster is more impressive than the so-called Jewel Box (NGC 4755), a brilliant, compact cluster in Crux lying some 6,400 light-years away.

Star clusters are spread here, there, and everywhere throughout the thin disk of the Milky Way. By contrast, nearly all the bright naked-eye stars we see in our sky are relatively close to us. We have already seen how close Alpha Centauri, Sirius, and Procyon are. Canopus, the brightest star in the deep southern sky, is 74 light-years distant. Arcturus, the brightest star in Boötes, lies 34 light-years away, while nearby Vega in Lyra is some 25 light-years distant. In Orion, Betelgeuse and Rigel are about 1,400 light-years away – anomalies due to their great intrinsic luminosities. Acrux, the brightest star in the Southern Cross, along with Antares in Scorpius, are each some 500 light-years distant. Deneb, the brightest star in Cygnus, is highly luminous and lies 1,500 light-years away. More typical are Regulus in Leo (69 light-years), Castor in Gemini (49 light-years), and Fomalhaut in Piscis Austrinus (22 light-years).

When we see planetary nebulae, we are, of course, foreshadowing the Sun's future when it runs out of nuclear fuel to fuse, some 5 billion years from now. The 3,000 or so planetary nebulae known in the Milky Way are scattered throughout the disk. The closest planetary to us, the Helix Nebula (NGC 7293), lies a mere 700 light-years away. The famous Ring Nebula (M57) in Lyra, one of the observational favorites, is 2,300 light-years distant. The Eskimo Nebula (NGC 2392) in Gemini lies at a distance of about 2,900 light-years. The Owl Nebula (M97) in Ursa Major is about 2,000 light-years away. The famous Dumbbell Nebula (M27) in Vulpecula is some 1,360 light-years distant. The Blinking Planetary (NGC 6826) is about 2,000 light-years away.

Measuring the distances to bright and dark nebulae is a trickier proposition, unless bright stars happen to be embedded in the nebulosity. Still, many of these objects are heartfelt favorites for backyard observers, and the several hundred well-known bright nebulae in the Milky Way, including emission nebulae (star forming regions) and supernova remnants, along with several hundred dark nebulae, form some of the more intriguing areas of our galaxy's disk.

Many of the most adored and observed objects in our galaxy have distances that are known relatively well, in large part because the star clusters associated with them have been well studied. The Orion Nebula (M42) lies 1,340 light-years from Earth. The Carina Nebula (NGC 3372) has a distance less well established, lying approximately 8,000 light-years from us. The Lagoon Nebula (M8) is some 5,000 light-years away, and the nearby Trifid Nebula (M20) is about 5,200 light-years distant. The Eagle Nebula (M16) lies some 7,000 light-years away. The North America Nebula (NGC 7000) is approximately 1,600 light-years distant. Dark nebula distances are tricky; the beloved Horsehead Nebula (B33) in Orion, as an example, is about 1,500 light-years away, whereas the Coalsack, in Crux, is some 600 light-years distant.

The most distant objects in the Milky Way that regularly attract astronomy enthusiasts armed with telescopes are globular star clusters, the immense spheres of old stars that inhabit our galaxy's halo. The well-known Hercules

Cluster (M13) lies about 22,000 light-years away, whereas the largest globular in the Milky Way, the southern sky's Omega Centauri (NGC 5139), is some 16,000 light-years distant. The brilliant southern cluster 47 Tucanae lies at a distance of 17,000 light-years. The condensed cluster M4 in Scorpius is 7,200 light-years away, while its neighbor M22 in Sagittarius lies at a distance of 11,000 light-years. The most distant globular cluster, NGC 2419 in Lynx, sometimes called the "Intergalactic Wanderer," is about 300,000 light-years away.

So just to revisit how large our Milky Way is, it would take that Voyager 1 spacecraft, traveling at 15 km/s, 1.8 billion years to travel from one end of the galaxy's disk to the other. And that's just our galaxy's bright disk. We have only begun to venture out into the deeper cosmos.

The Milky Way is but one of at least 54 galaxies that make up the so-called Local Group, our little family of spirals, dwarfs, and irregulars. The Local Group members are gravitationally bound as an assemblage of galaxies, and span about 10 million light-years across. That's 100 times the length of our galaxy's bright disk.

The Local Group is dominated by two spiral galaxies, the Milky Way and the Andromeda Galaxy. The Milky Way, as we have seen, contains some 400 billion stars in a disk that extends at least 100,000 light-years across, and our neighbor the Andromeda Galaxy is somewhat larger in size and is a spiral galaxy, some 2.5 million light-years from us, that is somewhat larger in diameter and contains as many as 1 trillion stars. But the Milky Way apparently contains much more dark matter than the Andromeda Galaxy, making the Milky Way perhaps slightly more massive than M31. As we have seen, in several billion years the Milky Way and Andromeda galaxies will merge because they are moving toward each other.

Another notable galaxy, the Pinwheel Galaxy (M33) in Triangulum, is a favorite for earthbound observers. Strangely enough, astronomers now know that M33 is a satellite of the Andromeda Galaxy and that its total luminosity is rivaled by the Milky Way's satellite, the Large Magellanic Cloud. The Pinwheel is a spiral whose bright disk spans some 70,000 light-years, and which lies some 2.7 million light-years away. It is the third wheel in the Local Group, after the Milky Way and Andromeda, and contains some 40 billion stars along with some very active star clouds and star forming regions.

The Local Group also contains some satellite galaxies of the Milky Way, most prominently the Large and Small Magellanic Clouds, beautiful features of the Southern Hemisphere sky, which appear as a detached, glowing portion of Milky Way star clouds. They are, in fact, irregular galaxies that orbit the Milky Way, and are strongly under our gravitational influence.

The LMC lies 163,000 light-years away and measures some 14,000 light-years across. It contains the famous Tarantula Nebula (NGC 2070), the largest and most active star-forming region in the Local Group. The SMC is some 197,000 light-years away and measures about 7,000 light-years across; it is a dwarf irregular galaxy. The Milky Way has 26 known satellite galaxies altogether,

Figure 12.4 The Milky Way and Andromeda galaxies are bound with dozens of other galaxies in what astronomers call the Local Group. The galaxy group moves together through space like one block in a vast cosmic city.
NASA/CXC/M. Weiss

most of which are tiny dwarfs, and undoubtedly most will eventually be gravitationally tugged into the galaxy and absorbed, as many have been in the past. The closest dwarf satellites of the Milky Way are the Canis Major Dwarf, which lies only 25,000 light-years away, and the Sagittarius Dwarf, about 65,000 light-years distant.

At large scales, of course, the universe is richly populated with groups and clusters of galaxies, although recall that the average density of the universe is very low – 1 hydrogen atom per cubic meter. Nonetheless, space is extraordinarily large, so lots of galaxies are out there even relatively close by. Beyond the Local Group, prominent groups of galaxies are many. They include the M81 Group, 11.4 million light-years away, which includes M81 and M82 in Ursa Major. The Centaurus A/M83 Group, at 11.9 million light-years, holds Centaurus A, M83, and NGC 4945, among others. The famous Sculptor Group, 12.7 million light-years distant, contains NGC 253, NGC 247, and others. The M94 Group, 13 million light-years away, holds M94 and others. The M101 Group, at

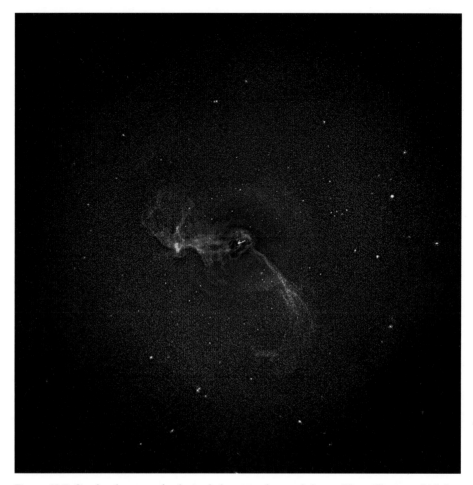

Figure 12.5 Our local group of galaxies belongs to the much larger Virgo Cluster, which is made up of thousands of galaxies. It is centered more than 50 million light-years away toward the constellation Virgo.
NASA/CXC/Stanford/I. Zhuravleva *et al.*

a distance of 24 million light-years, has M101 and half a dozen others. And it goes on and on and on.

Finally, when we move out to a distance of about 54 million light-years, we come to the closest big cluster of galaxies, the Virgo Cluster. Dozens of members of this rich cluster of galaxies can be observed in backyard telescopes, and they include the supergiant elliptical M87, elliptical galaxies M84 and M86, the beautiful spiral M100, and many more exceptional objects, including Messier catalog standouts M49, M58, M59, M60, M61, M85, M88, M90, M91, M98, and M99. This huge cluster, containing about 1,300 large galaxies spread over a diameter of more than 11 million light-years (and many more dwarfs), forms the heart of a larger agglomeration of galaxies called the Virgo Supercluster of galaxies, of which we are a part. The Virgo Supercluster, the largest collection of galaxies in our region of the universe, spans some 100 million light-years and contains more than 100 galaxy groups and clusters.

The diameter of the Virgo Supercluster, then, is some 1,000 times larger than the diameter of the Milky Way. And yet, when we look at the incredible immensity of the Virgo Supercluster, the size of the visible cosmos is 1,000 times larger yet. The universe, then, is at least a million times larger than the diameter of the Milky Way Galaxy. The difference in volume, however, is far more impressive. The volume of the Milky Way is approximately 10^{60} cubic meters. But the volume of the observable universe is at least 10^{80} cubic meters, which allows all of those 100 billion galaxies to comfortably fit within the cosmos.

Beyond the Virgo Cluster lie numerous clusters and superclusters of galaxies, which taken together are termed the large-scale structure of the universe. The most remote mature cluster of galaxies yet discovered, which was found in 2011, is CL J1449+0856, which has a redshift of 2.0, making it more than 32 billion light-years distant. The Hubble Ultra Deep Field, a small region of the constellation Fornax, has been studied by astronomers using the Hubble Space Telescope beginning in 2003. It contains protogalaxies that are some 13 billion years old, showing them as they were 800 million years after the Big Bang.

For many years, sky surveys of various types have constructed models and maps of the very distant universe, painting a picture of the large-scale structure. Beyond superclusters of galaxies, larger structures appear to be arrayed in sheets, walls, and filaments winding across the cosmos. Between these sheets and walls lie sparsely populated areas of galaxy clusters and superclusters called voids. The overall effect on very large scales is something akin to a foam with cavities, or a three-dimensional web with voids, and the earliest of these concentrations was discovered in 1983. The past 30 years has seen an explosion in understanding of the cosmos at very large scales.

Canadian astronomer R. Brent Tully (1943–) identified the huge filament that contains the Milky Way (and the Virgo Supercluster), called the Pisces-Cetus Supercluster Complex, in 1987. This mammoth structure stretches some 1 billion light-years across. Also in 1987 astronomers found a huge cavity in the universe's structure they dubbed the Giant Void. In 1989, astronomers at the Harvard-Smithsonian Center for Astrophysics, including Margaret Geller (1947–) and John P. Huchra (1948–2010), found evidence of the so-called Great Wall, a sheet of galaxies measuring 500 million by 200 million by 15 million light-years. Astronomers using the Sloan Digital Sky Survey data discovered another mammoth structure in 2003, the Sloan Great Wall. In the Hydra-Centaurus Supercluster, a feature called the Great Attractor, some 200 million light-years from us, is pulling galaxies in toward it.

Other studies have uncovered still more large structures and voids in the big universe. The most distant individual object for which good data exist appears to be a galaxy candidate known as UDFj–39546284, associated with a gamma ray burst, whose redshift would place it some 40 billion light-years distant, meaning that the burst happened a mere 630 million years after the Big Bang.

The techniques astronomers use to measure distances in the universe are nearly as varied as the inhabitants of the universe. We have already discussed the technique of parallax, of observing objects at different times of the year when Earth is placed differently in its orbit. This allows the measurement of a shift in positions of stars and other objects relatively close by, as compared with extremely distant objects in the background.

Direct measurements of distance are tricky, and generally are reliable out to about 3,000 light-years – only 3 percent of the width of the bright disk of our galaxy, for example. To measure larger distances, astronomers use an array of objects known as standard candles – that is, objects of precisely known intrinsic brightness whose apparent magnitude can then be used to derive a distance. But they need to be careful with calibrations of standard candles, and also to be sure they are recognizing the distant object properly.

In our galaxy, dynamical parallax uses the orbiting stars of a binary system, whose masses, sizes of orbits, and orbital periods can be used to derive a distance. Large telescopes can also measure the properties of eclipsing binary stars. This enables astronomers to measure distances to them, and this includes eclipsing binaries as far away as those in nearby galaxies in the Local Group. A specific type of variable star, so-called Cepheid variable stars, have a period–luminosity relation that allows them to be used to estimate distances because the relationship between their periods and absolute magnitudes is so well known. For use with old stars, such as in globular star clusters or nearby galaxies, astronomers have several tools; variable stars called RR Lyrae stars as well as specialized techniques such as the tip of the red giant branch in the cluster's Color-Magnitude Diagram, the planetary nebula luminosity function, the globular cluster's overall luminosity function, and surface brightness fluctuation.

X-ray bursts can also be used as standard candles, as can interstellar masers (used essentially as precision parallaxes, and which have provided direct distances to Local Group galaxies M33 and IC 10), Cepheid variable stars (they were used by Hubble to identify the Andromeda Galaxy as a distant object!), and novae.

At greater distances, studies of the properties of individual galaxies within galaxy clusters provide a helpful estimate of distance. Further, the so-called Tully–Fisher relation, developed in 1977 by R. Brent Tully and his colleague, American astronomer J. Richard Fisher (1943–), is very useful. The Tully–Fisher relation stipulates an empirical relationship between the intrinsic luminosity of a spiral galaxy and its velocity width, which is the amplitude of its rotation curve.

Additionally, the Faber–Jackson relation, developed by American astronomers Sandra M. Faber (1944–) and Robert E. Jackson (1949–), is helpful with estimating distances to elliptical galaxies. This says the luminosity of a elliptical galaxy relates to its stellar velocity dispersion, a statistical dispersion of velocities within the galaxy.

Perhaps the most important standard candles of recent decades are Type Ia supernovae, which are exceedingly well known and have a systematic relation between absolute brightness and period, allowing them to be a highly reliable source of distances to galaxies. They result from the catastrophic explosion of a white dwarf in a binary system that is drawing matter from a companion, reaching a critical mass and setting off an explosion like a cosmic bomb. They were the lynchpin behind the evidence for dark energy, in fact, as well – something we will come to in a later chapter.

And redshifts can also be great distance indicators to distant objects, although they are dependent on certain assumptions. Fortunately, data from the Hubble Space Telescope, WMAP, Planck, and other instruments have dramatically reduced the uncertainties of various cosmological parameters, making galaxy redshift measurements an increasingly reliable way to estimate big distances in the universe. There is a limitation here, too – the very distant universe is fainter and also different from where we are, in that time appears to advance more slowly than it does on Earth. So astronomers see the distant past both redshifted and time dilated, which makes studies of very high redshift objects extremely difficult.

We know it is a vast cosmos filled with innumerable objects, all of which we want to measure the distances to. In many cases, the results are getting far better, and the distance scale is coming into clearer focus. But there are still significant limitations in seeing the universe clearly.

And one of the most intriguing comes from the ages-old subject of dark matter.

Chapter 13
The mystery of dark matter

Despite the incredible advances in so many areas of astronomy over the past generation, some things remain works in progress. One of these is the mystery of dark matter, one of the fundamental components of the universe. Cosmological results in 2013 made by the Planck spacecraft team suggest the newest breakdown of the composition (mass-energy) of the universe as 4.9 percent ordinary or baryonic matter, 26.8 percent dark matter, and 68.3 percent dark energy. The story of the nearly opaque mystery of dark energy follows in the next chapter. For now, we'll explore the strange stuff that cosmologists know exists but the nature of which remains murky – dark matter.

In the early part of the twentieth century, not long after the basic cosmic distance scale and the nature of galaxies as separate "island universes" was discovered, astronomers began to stumble on clues from several directions that the bright stuff they saw in the cosmos wasn't the whole story. In 1932, Dutch astronomer Jan H. Oort (1900–1992), later to become famous for his hypothesized cloud of comets surrounding the solar system, was busily studying the motions of stars in the Sun's neighborhood. He determined that the mass of the Milky Way must amount to more than the luminous disk, the halo, and globular clusters. But a short time later, astronomers had serious doubts about the measurements.

Soon thereafter, Swiss astronomer Fritz Zwicky (1898–1974), a colorful, clever, and cantankerous fellow, took up the problem, and Zwicky was an amazing thinker and true innovator. While at the California Institute of Technology, Zwicky studied several clusters of galaxies and found that their masses must have been greater than the visible light they emitted could account for.

In a famous paper published in 1937, Zwicky suggested the possible existence of dark matter, calling it *dunkle Materie*, and shared the observations he had made over previous years of a variety of galaxy clusters. He didn't really push the idea, but raised it as one possible solution to the observations.

In the grand tradition of nineteenth-century science, Zwicky had become the first astronomer to widely observe and analyze a range of clusters of galaxies, cataloguing them, studying their compositions, and suggesting that some unseen material must be present to explain the orbits of the individual galaxy members, which without the unseen material would fly off into intergalactic space.

In fact, Zwicky found that gravity alone would be only $1/100^{th}$ as powerful as the force that would be required to hold the galaxies together in these clusters. Something else, something that couldn't be seen, might be acting as a "glue" in the equation, or else the observations would not make sense at all.

And yet Zwicky's work, convincing on the face of it, would be viewed mainly with skepticism until many years later. Zwicky was a visionary who foretold solutions to several big riddles of astrophysics, but his reputation as a cranky guy helped to put people off from seeing his brilliance. This, despite the fact that he had realized – thanks to observations made by Walter Baade – that supernovae could be used to measure distances in the cosmos; that dense neutron stars would result from supernova explosions; and that galaxies could act as lenses that produce images of distant background objects.

His seminal 1937 paper, *On the Masses of Nebulae and Clusters of Nebulae*, might have made a spectacular splash. But other astronomers were wary of Fritz Zwicky, and the results languished for decades. Other astronomers worked on galaxies, galaxy clusters, and their masses, but no one got traction on the idea of dark matter playing a huge role in this business. Then came a period with relatively slow progress. Astronomers made progress with the Milky Way's rotation curve and the bulk motions of galaxies. But relatively little progress took place until the early 1970s.

Then along came another influential paper. In 1974, American astronomers Jeremiah P. Ostriker (1937–) and Amos Yahil (1943–), and Canadian–American physicist James E. Peebles (1935–) published a study that assembled the myriad clues about dark matter into a coherent proposal, that galaxies must have masses 10 times greater than the visible light they emit could account for. They proposed, as a few others had thought, that this dark matter exists in giant halos surrounding the disks or ellipses of galaxies. They supported this notion by analyzing rotation curves of galaxies, showing that the galaxies' disks rotated as fast far away from the center as they did closer in, which suggested some unseen material must account for an increase in mass away from the galactic centers.

The team also showed, from studying the motions of the galaxies, that their masses must be great due to the unseen material detected. The gravitational attraction between the galaxies is therefore so great that it has reversed the natural motion of all galaxies to move away from each other, and caused them to be on a collision course. The cluster potential is also partly responsible, as other clusters are gravitationally bound. In the last generation, oodles of evidence of collisions between galaxies in groups and clusters has come to light.

The Ostriker-led team found that for the Milky Way, the galaxy's mass enclosed within a sphere follows a simple rule, that it is proportional to the radius of the sphere out to about 1 million light-years, where it reaches a value of 1 trillion solar masses. This result conflicts strongly with images of galaxies, whose light falls off sharply away from the galactic center. Most of the light in a photograph of a Milky Way-sized galaxy darkens into nothingness as you move about 50,000 light-years away from the center. But this result showed that most of a galaxy's mass lies more than 300,000 light-years from the center, where the galaxies we see in the cosmos appear to be dark. What exactly was going on here?

This work was based on spiral galaxies, but astronomers know they are a minority among types of galaxies. What about elliptical galaxies? American astronomer Herbert J. Rood investigated elliptical galaxies and found the mass to light ratios in these galaxies to be even larger: they may contain as much as 200 times dark matter as bright matter. So a real mystery sprang forth in the early 1970s. Astronomers began to call this the puzzle of the "missing matter," or simply dark matter.

The work on observing the rotational characteristics of galaxies carried on in the late 1970s and 1980s. American astronomers Vera Rubin (1928–) and W. Kent Ford, Jr. (1931–) took a leading role in producing many papers focused on individual galaxies, which helped guide the analysis of the role of dark matter in the universe. Rubin had also been a pioneer in the field in the 1950s, making early measurements of the rotation curves of galaxies.

Their group, centered at the Carnegie Institution of Washington, was the first to systematically use the new technologies of image-tube spectrographs to collect large numbers of observations. This enabled them to collect rotational velocities for the disks of galaxies that extended farther out from the galactic centers than had ever been seen before. The evidence was consistent with earlier observations, but now was accumulating in large volumes, and went beyond mere visual observations to become convincing to most astronomers and physicists.

By about 1980, then, astronomers had accepted that the universe contained large amounts of some kind of dark matter. The motions of individual stars and gas clouds in galaxies betrayed its presence. The motions of galaxies within clusters gave it away. And it was a significant, even a shocking component of the cosmos: the evidence suggested that some dark component of the universe had 10 times the mass of the stars observed by astronomers, and some 5 times more mass than all of the ordinary matter of the universe combined.

The research by Rubin and Ford and all of their colleagues and predecessors suggested that this dark component of the universe was not distributed randomly, either. It seemed to exist in the general regions of concentrations of stars, but was much more widely distributed than stars, and clearly seemed to exist in enormous halos surrounding galaxies.

Some scientists claimed that perhaps dark matter was a flawed concept, and that something was different about gravity than the way we understood it. But multiple lines of evidence for dark matter began to emerge through a huge variety of research projects. The effects of dark matter can be observed directly

Figure 13.1 An immense gravitational field surrounds galaxy cluster Abell 68 and acts as a magnifying lens for astronomers. It allows them to study galaxies aligned behind the cluster and see them as they were billions of years ago, despite the incredible faintness of their light. NASA and ESA

"in action," as with the Bullet Cluster, a pair of colliding clusters of galaxies lying some 3.7 billion light-years away in the constellation Carina.

In this system, the two clusters of galaxies have collided and the stars in the galaxies have essentially passed through each other, while the more densely arranged atoms of gas have collided and pulled out of the stellar systems – this is what astronomers observe in the Bullet Cluster. By examining weak gravitational lensing of background galaxies, astronomers can see a concentration of mass toward the centers of galaxy clusters like the Bullet Cluster. Gravitational lensing had been predicted by Fritz Zwicky all the way back in 1937. Gravitational lensing occurs when a galaxy cluster, for example, acts as a lens to beam light from an object lying directly behind it, creating images of the distant object that can be seen around the cluster. The principle is explained from Einstein's General Theory of Relativity.

Zwicky explained that the separation of images produced by clusters of galaxies would mean images from beyond would be clearly visible around the galaxy clusters, and not obscured by the clusters themselves. And lo and behold, instruments like the Hubble Space Telescope have shown since the 1990s multiple examples of "arcs" and other features around clusters of galaxies, produced by a gravitational lensing effect. (In some cases, this had been shown before Hubble.) And the effect is exactly consistent with having been produced by dark matter halos.

The evolutionary growth of structures in the universe also supports the existence of dark matter. The current universe astronomers are familiar with is lumpy: it has stars, galaxies, clusters of galaxies, gas clouds, and voids that appear to be underdense but are not empty. Measurements of the cosmic background radiation suggest that the Big Bang produced a universe that was very smooth, at least initially. So how could the cosmos have gone from very uniform and smooth to lumpy and not uniform at all? Astronomers hypothesized that in the early universe, ripples must have existed from which the universe's later lumpiness grew.

For years, astronomers had unsuccessfully looked for anisotropies in the cosmic microwave background radiation, but always unsuccessfully. It became clear that something like cold dark matter was needed to produce the observed large-scale structure in the universe, but no one had yet detected evidence of it.

In the 1990s, astronomers began to have success looking for evidence of these seeds with the Cosmic Background Explorer (COBE) satellite. Indeed, the COBE team, led by American astronomers George Smoot (1945–) and John C. Mather (1946–), found evidence of the seeds in their first release of data. Observed fluctuations in the cosmic microwave background were small, about 1 part in 100,000, but they were there, and could have grown into non-uniformities that led to stars, galaxies, and clusters of galaxies. But they could not have done so if what we see as bright matter is all that exists. The solution requires a significant amount of dark matter to produce the universe we see around us.

Dark matter fluctuations in the early universe would have exerted enough extra gravitational force to allow the non-uniformities to grow, and calculations demonstrated the amount of dark matter needed to accomplish this matched the amount inferred to exist from observations of galaxies and galaxy clusters – 5 to 10 times that of bright matter.

Later observations by other cosmology satellites refined the picture. A joint project of NASA and Princeton University, the Wilkinson Microwave Anisotropy Probe (WMAP) was launched in 2001 and studied the cosmic microwave background intensely for nearly a decade, playing a key role in establishing the current understanding of dark matter and the standard model of cosmology. In 2009, the European Space Agency launched the Planck satellite, which has delivered the best-yet cosmological results relating to the composition and age of the universe and clues about the nature of dark matter.

But let's return to that key period, around 1980, when astronomers were solidifying their evidence of dark matter's existence. Not only did the astronomy world begin to embrace and accept the reality of dark matter during that time, but also a majority of scientists began to lean toward suspecting dark matter consisted of some type of subatomic particles. The idea was that perhaps some type of dark – nonbaryonic – particles may exist that interact only weakly with baryons and photons. This was a radical idea in the early 1980s, but now it is taken as the norm.

At this juncture, the existence and nature of dark matter ties back into cosmological models, as we explored in the Big Bang chapter. Astronomers had long believed that the matter distribution in the universe, as we have explored, could only be explained by fluctuations in the early universe's density, perhaps 10^{-34} seconds after the Big Bang. Before the COBE results, before WMAP, before Planck, no one knew whether these fluctuations existed and whether any evidence would be found for them. They did know that enormous numbers of cosmological relic neutrinos existed from the early universe, and that weak interactions can convert these neutrinos to photons, and vice versa. But as the universe expanded, they calculated, the neutrinos' ability to change "froze out" and so they persisted as relics of that very early age. In the present universe, every cubic centimeter of space contains roughly 113 neutrinos of each of the three types – more than 300 neutrinos altogether.

A generation ago some astronomers believed that neutrinos might have a small, non-zero rest mass, and that this may help to solve the dark matter problem. Some cosmologists also expanded this notion to explore whether other, as-yet undiscovered relic particles of the Big Bang could also be part of the answer for dark matter.

In the early 1980s, a number of cosmologists at different institutions all came to a similar realization at nearly the same time. The idea that non-baryonic dark matter in the form of neutrinos or more massive, unknown particles could resolve the differences between theory and observations was a powerful one. They all discussed the possibility that a non-interacting dark matter "fluid" of particles very early in its history could create fluctuations that would lead to the observed matter distribution we see in the universe much later on.

This process of thinking about dark matter in the 1970s and 1980s underscores an incredible coincidence – that so much thinking about dark matter and its relationship to galaxies and clusters of galaxies was followed, a short few years later, with a wave of thinking about dark matter in terms of subatomic particles being necessary to help explain the main cosmological models everyone had come to accept. In the minds of many astronomers and cosmologists, this period was one dominated by dark matter!

These were important results – neutrinos could account for at least some of the observed dark matter, they exist in the universe in numbers comparable to cosmic microwave background photons, and they do have a very small mass. In 2014, physicists reported the best-yet measurements of neutrino mass, with the

Figure 13.2 Increasingly, sensitive dark matter detectors have come online in recent years, as more scientists get involved with the hunt for the missing universe. In Lead, South Dakota, physicists are using a former mine, the famous Homestake Gold Mine, to protect their Large Underground Xenon experiment from unwanted cosmic radiation that constantly bombards Earth.
Sanford Lab

sum of the masses of the three flavors of neutrinos as 0.320 ± 0.081 eV. Even before accurate measurements of neutrino mass, in around 1980, the possibilities of neutrinos being responsible for the universe's dark matter loomed large.

But then astronomers encountered stumbling blocks. First, they began thinking about structures in the universe that exist on relatively small scales, like galaxies. These structures formed relatively late in the evolutionary history of the cosmos. Particles like neutrinos are relativistic – that is, they move at nearly the speed of light and freely stream from horizon to horizon. Cosmologists term dark matter that moves at such speeds when it decouples from photons "hot dark matter."

This ties back to those density fluctuations in the early history of the universe. Hot dark matter like freshly decoupled neutrinos streams freely at relativistic speeds, erasing the density fluctuations and preventing them from living on to be the "seeds" that promote the formation of high densities of matter like stars, galaxies, and galaxy clusters.

In a universe controlled by hot dark matter, cosmologists can calculate that neutrinos would become non-relativistic when the universe was about 700 years old and 700 light-years in diameter, a sphere that by now would have expanded to some 60 million light-years. In this scenario, the lowest-mass objects that could begin gravitational collapse, when matter begins to dominate, would be superclusters of galaxies. This would suggest a "top-down model" of the universe, in which huge things form first, followed by galaxies and then stars and smaller bits of matter.

Also around that magical era of 1980, cosmologists began to simulate the early history of the universe as portrayed by hot dark matter models using sophisticated computers with huge numbers of particles. Repeatedly, simulations revealed that a hot dark matter model had galaxies forming much later in the cosmos than expected; in fact, galaxies were observed significantly farther back in time than the hot dark matter model of cosmology predicted. This was almost immediately a major problem for hot dark matter cosmology.

Another significant problem arose with the hot dark matter model when, in 1979, American astronomer James Gunn (1938–) and Canadian astronomer Scott Tremaine (1950–) produced an influential paper describing the limit to the number of neutrinos that can exist in a galaxy's halo. They argued that the limit depends critically on a neutrino's mass. Their calculations showed that if neutrinos were to form a substantially massive galaxy halo, then their mass must be greater than 30 eV, a figure that even then no one believed was possible. So the consensus was that neutrinos could not form the halos of low-mass galaxies.

Not only did this paper throw a kink into the momentum of believing that neutrinos played a big role in the halos of galaxies, but it also seemed to halt the enthusiasm for neutrinos playing a substantial role in dark matter in any place in the cosmos. And thus the enthusiasm for a hot dark matter model of cosmology began to wane. But there was a major alternative waiting in the wings.

In 1982, Peebles was among the first to propose it, the concept called cold dark matter, or CDM. Within 2 years a number of influential researchers produced papers analyzing this alternative, and they included American astronomers George Blumenthal (1945–), Heinz Pagels (1939–1988), Joel Primack (1945–), and Sandra Faber (1944–), and English astronomer Martin Rees (1942–).

The main property of cold dark matter is that it consists of particles that are traveling at speeds slower than relativistic velocities when they decouple from photons. Cosmologists proposed that such particles, which could have a mass of 100 billion eV that decouples when the universe's temperature is 10 billion eV, would be moving at speeds of much less than that of light. And correspondingly, these cold dark matter particles would not obliterate the density fluctuations in the early universe, such that the early ripples that would lead to structures in the cosmos would be preserved. And the CDM models suggested that fluctuations in the energy density of the cosmos could exist and remain throughout this process at various scales, meaning that instabilities could grow throughout the CDM fluid and become galaxies and clusters of galaxies as the universe aged.

As the 1980s progressed, more and more astronomers and cosmologists gained confidence in the CDM models. These data proposed that neither of the two major problems encountered with hot dark matter existed – galaxies could form early in the CDM simulations, exactly as they are observed in the real universe, and because the dark matter is cold, in other words moving relatively slowly, it can exist in copious quantities in galaxy halos, as the observations suggest. The next challenge resulted from the question many scientists had hoped to understand ever since the days of Fritz Zwicky: What exactly is dark matter made of?

Cosmologists immediately turned to subatomic particles as the likeliest culprits of supplying dark matter. The so-called standard model of particle physics has been extremely successful at explaining the behavior of matter, and also in predicting new discoveries of more exotic particles over time. The known subatomic particles include fermions and bosons, which are characterized by different spins, or angular momentum. Fermions include electrons and baryons, which include protons and neutrons. Electrons cannot be further divided, but baryons can – they are made of quarks. Electrons belong to a class of fermions called leptons, and among leptons there are not only electrons but also muons and tau particles. Electrons have their anti-particle, a positron, while anti-muons and anti-tau particles also exist. Each charged lepton also has an associated neutral particle, the neutrino. Three "flavors" of neutrinos exist, the electron neutrino, muon neutrino, and tau neutrino.

Three groups of two quarks also exist, with the lowest mass, most stable quarks comprising baryonic matter, the up quarks and down quarks. The strong force holds quarks together, and three of them comprise a baryon. The fractional electrical charges of quarks come together so that two up quarks and one down quark held together by the strong force comprise a proton with an electrical charge of 1.

Bosons are the carrying particles of nature that mediate the strong force that binds quarks together (gluons), mediate the weak force (W and Z bosons), carry the electromagnetic force (photons), carry the gravitational force (gravitons), and generate the mass of particles (the Higgs boson).

Cosmologists know that if subatomic particles are responsible for dark matter, the particle in question would not be charged, or it would be detectable through its electromagnetic interactions. This rules out electrons and protons as well as muons, tau particles, W bosons, and Z bosons, and some types of pions and other mesons/hadrons.

Because dark matter particles are long lived, the particle in question should be highly stable. This eliminates neutrons, tau particles, muons, W and Z bosons, the Higgs boson, pions, and other types of mesons and hadrons (except for protons). Dark matter particles, of course, have to have energy. In the world of particle physics, this leaves just one potential culprit, neutrinos. Cosmologists know that neutrinos exist in huge quantities; they have mass, and they are electrically neutral. But they have already been eliminated as the chief

Figure 13.3 The $2 billion Alpha Magnetic Spectrometer onboard the International Space Station is being used to search for hints of dark matter particles that might be blocked out by Earth's atmosphere.
NASA

component of dark matter, so the CDM model suggests that dark matter must be made of something other than a citizen of the realm of particle physics.

However, the standard model of particle physics may not be the end state of reality. Some phenomena in the cosmos are not explained in the context of the standard model. The observed masses of neutrinos do not fit the standard model very well. Could another model of particle physics be the reality, such as the leading candidate, supersymmetry?

Supersymmetry postulates that symmetry exists between fermions and bosons; in other words, that every boson has a fermion partner with half integer spin and every fermion has a boson partner with integer spin, so that every known particle has a supersymmetric partner. Electrons would have selectrons, muons smuons, tau particles stau particles, neutrinos sneutrinos, quarks squarks, photons photinos, W bosons Winos, Z bosons Zinos, gluons gluinos, gravitons gravitinos, and Higgs bosons Higgsinos.

Doubling the number of particles in the universe in this way might in part resolve the neutrino mass and other problems of the standard model, but is it reality? None of these hypothetical supersymmetric particles has been found. It is possible that when the universe was young it contained huge numbers of these supersymmetric particles, almost all of which have decayed over the past several billion years, leaving very few at the present day. Cosmologists suggest that perhaps only the lowest mass supersymmetric particle cannot decay and is therefore around still, and perhaps it is the dark matter culprit. They call it the lightest supersymmetric particle, or LSP.

But the nature and existence of the LSP is, of course, speculative. The CDM model of cosmology emerged intact, however, despite lots of other possibilities of what might constitute dark matter in the universe. It suggested a cosmos of perhaps 5 percent baryonic matter, maybe 10 percent of this visible as bright matter, and the rest of the matter in the universe consisting of some sort of dark material. Another hypothetical candidate that emerged in the 1980s was weakly interacting massive particles, or WIMPs, which refers to a dark matter particle that arose from the hot, dense plasma of the early cosmos. Strong support for the existence of WIMPs comes from the abundance of dark matter in the universe today matching the product of what cosmologists would expect from WIMPs being produced in the range of 100 GeV in mass. The existence of WIMPs would agree with supersymmetric ideas as well. But recent experiments have failed to detect WIMPs, and supersymmetry has not received any experimental support from the Large Hadron Collider either.

What does the emerging model of CDM say about how galaxies, stars, and other objects form in the universe? Cosmologists believe that CDM halos form by gravitational instabilities in the expanding universe. These halos consist of dark matter particles, they reason, that follow only gravity. The structures of dark matter and bright matter that observers were finding in the 1980s and 1990s went hand in hand with discoveries of the large-scale structure of the universe, the spider-web nature of galaxy superclusters set apart by large cosmic voids. This weblike structure of the cosmos agreed with what CDM would predict, which was comforting to cosmologists and astronomers alike.

In the mid-1990s, a group of researchers led by Argentinian astrophysicist Julio Navarro (1962–), Mexican–British cosmologist Carlos Frenk (1951–), and English astrophysicist Simon White (1951–) worked out the details of how galaxy halos form by gravitational collapse in a CDM universe. They found that halos in galaxies have a similar density distribution, from small galaxies all the way up to huge superclusters. This finding, called the NFW profile, showed that the density of dark matter increases right into the centers of galaxies and clusters, and falls off rapidly at large radii.

But low mass, low surface brightness galaxies offer a bit of a problem for the CDM model. American astronomer Stacy McGaugh has shown that some low-mass galaxies do not fit the NFW profile; but the differences here may be caused by the CDM model focusing on the cosmological formation of galaxies, not taking into account unusual examples of galaxy evolution.

And just when astronomers thought they had enough to contend with in deciphering and testing the CDM model, in 1998, along came dark energy. This will be treated in full in the next chapter. But suffice it to say that the discovery of an effect that is accelerating the expansion of the universe sent cosmology into brand new levels of complexity. The dark energy discovery reintroduced

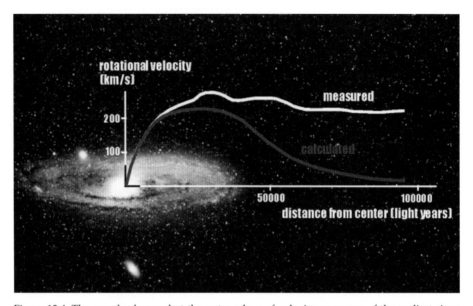

Figure 13.4 The speeds observed at the outer edges of galaxies were one of the earliest signs that something was wrong. Astronomers now know that the disks are rotating faster than expected because most of the galaxy's mass is hidden as dark matter.
Queens University

into the equation Einstein's cosmological constant, which he introduced in relativity, abandoned, and gave up on deciphering. The cosmological constant essentially introduces a repulsive force into the universe. This repulsive force apparently dominates and leads to the acceleration of the expansion of the universe as observed in the dark energy discovery observations.

The cosmological constant can be thought of as a dense fluid that has negative pressure and does not dilute as the cosmos expands. This constant fluid still affects the universe fundamentally. Suddenly, the discovery of dark energy cast aside some of the shortcomings of the standard CDM model. Around the time of the dark energy discovery, Ostriker and his colleague, American cosmologist Paul Steinhardt (1952–), were proposing a "concordance model" of the universe in which a cosmological constant was behind the cosmic expansion. As the new millennium turned, cosmology entered a new, alien world: dark energy seemed real, and the standard CDM model of the universe was cast asunder. Now, cosmologists turned to the Lambda-CDM model, which incorporates the observations and theory of a dark energy-dominated universe.

In the new model, Lambda stands for the cosmological constant. Cold dark matter is still believed to play a huge role in the history of the universe. But what is it? We still don't know. Most of the leading possibilities at the particle and subatomic particle level have not as yet panned out.

CDM is still hypothetical, although it is absolutely the leading idea for explaining key elements of the universe. Since its rise in 1984, the objects making up CDM have been suggested to be of several types. Cosmologists have proposed WIMPs, but they are yet to be discovered or perhaps do not exist. They have also proposed that axions make up dark matter or a significant part of it. Axions are hypothetical elementary particles that were proposed in 1977, just a short time before CDM became the dominant model. They have a very small mass and, if they exist, hypothetically would interact with matter in the right way to explain many of the properties of dark matter.

Alternatively, for a time, some cosmologists had suggested that dark matter is composed of MACHOs – massive compact halo objects. These are not from the world of elementary particles, but rather are large objects potentially consisting across an entire spectrum of fascinating astronomical objects. It is possible that vast numbers of black holes, neutron stars, white dwarfs, or free-floating planets contribute significant amounts of mass in galaxies and galaxy halos, and compose at least some of the dark matter. But the bottom line is that MACHOs have been ruled out in the minds of nearly all cosmologists, for a variety of reasons.

Nearly 80 years after Fritz Zwicky raised the existence of dark matter, we are still looking for the answers. But the search has led to a solid understanding of much of galaxy dynamics, and a highly promising cosmological model. And it has also led us into the very strange world of dark energy.

Chapter 14
The bigger mystery of dark energy

As we touched on in the last chapter, the apple cart of cosmology was upset in a brutal and surprising way in 1998, with the observations of distant supernovae, exploding stars, by two big teams of research astronomers. Their discovery revealed the mysterious force that is accelerating the expansion of the cosmos and came to be known as dark energy. This is a crucially important area of cosmology and astrophysics, as recent observational results suggest that approximately two-thirds of the energy/mass of the universe consists of dark energy. So for the last 17 years, astronomers have realized they know very little about what most of the universe is composed of.

The discovery took the astronomy world by storm. But the seeds of the dark energy idea began much farther back in time, to be precise in 1915, actually a century ago. Prior to this, the universe was one of mechanical physics, of Isaac Newton's absolute, predictable cosmos that worked like a clock, independent of other factors. In 1905, German physicist Albert Einstein (1879–1955) began to overturn this, with his Special Theory of Relativity. In 1915, Einstein, late in the year, presented his General Theory of Relativity to an audience in Berlin. Einstein, driven in something of a race with potential competitors, unveiled his equations that not only introduced relativity but also described how space-time is distorted by mass, how mass moves around in gravitational fields, and the many forms of motion in the cosmos as observed by different frames of reference.

Einstein freed the world from seeing the universe in a Newtonian way, a cosmos beset by hard, unchanging geometry and now transformed from fixed coordinate grids to a changing, dynamic system in which space and time are linked and transform in various ways as codependents. That most important Einsteinian equation, $E=mc^2$, was merely the simplest, and it did indeed reveal that matter and energy in the cosmos are coequals, the same thing, in different forms, and that matter can be converted into energy and vice versa. (Your metabolizing this morning's breakfast is one proof of this.) But Einstein provided many other field equations that supported his theories, and he did not ultimately solve them all.

Figure 14.1 Massive galaxy clusters like Abell 1689 allow astronomers to see the strange effects of dark energy playing out on a cosmic scale.
NASA, ESA, the Hubble Heritage Team (STScI/AURA), J. Blakeslee (NRC Herzberg Astrophysics Program, Dominion Astrophysical Observatory), and H. Ford (JHU)

One of those who took up the challenge of investigating relativistic equations was Einstein's correspondent, German physicist Karl Schwarzschild (1873–1916). Despite the fact that he was stationed on military service at the Russian front in World War I, Schwarzschild tinkered with Einstein's equations and found that if matter was compressed enough, the equations would all mean that space-time would be torn asunder, leading to a black hole – see the next chapter. At the same time Schwarzschild was producing intriguing new results from Einstein's equations, he contracted a rare autoimmune disease, which led to his death at the age of just 42 years.

Having lost his best consultant, Einstein began a vigorous communication with Dutch physicist Willem de Sitter (1872–1934) and Austrian–Dutch physicist Paul Ehrenfest (1880–1933). Orbits calculated from the Einstein–

Schwarzschild solutions were very slightly different than those predicted by Newtonian physics, and explained an anomaly noticed by observers of Mercury's orbit many years earlier. Relativity, it seemed, was onto something and explained the universe in a better and more sophisticated way than the convention.

Continuing his correspondence with his Dutch friends, Einstein proposed an idea that at first he thought was absurd: a finite universe unbounded in space-time, the three-dimensional equivalent of traveling around a planet forever and ever without reaching an edge. No one had proposed such a thing before. The universal ideas previously had been a static, infinite universe or a finite universe placed centrally in space, neither of which would work with general relativity. With this new idea, Einstein reasoned, masses in space curve, and with enough mass, the entire universe could curve back on itself, although observers traveling through it would appear to themselves to be traveling in straight-line paths.

Thus, in Einstein's new universe, the age-old question everyone wants to know, "What lies beyond the edge of space?," has no meaning, as there is no edge, no wall, no limit. But Einstein's new concept had a problem. The idea required that the universe be expanding or contracting in order to make sense – the universe could not be static. And in 1917, no one yet knew that the cosmos was expanding. (At Lowell Observatory in Flagstaff, Arizona, American astronomer Vesto M. Slipher [1875–1969] was beginning his work on this field, but had not yet accumulated or published significant results.)

Einstein changed his equations to allow a static universe to exist within the framework. He added a mathematical figure that represented balancing gravity on large scales, and called it the cosmological constant, denoted by a Greek lambda, Λ. This "fudge factor" introduced by Einstein was described by him as "necessary only for the purpose of making possible a quasi-static distribution of matter, as required by the fact of the small velocities of the stars."

Einstein's introduction of the cosmological constant was a flawed move, motivated by a desire to have a static universe, and, of course, a dozen years afterward astronomers discovered that the universe isn't static. Work on the redshifts of galaxies leading up to the early 1920s was accomplished by Slipher and by fellow American astronomers William W. Campbell (1862–1938), Edwin P. Hubble (1889–1953), and Milton Humason (1891–1972). Strangely, Einstein was slow to accept the work of these observational astronomers for the expanding universe.

He did come to regret adding the cosmological constant, this anti-gravity force, to the equations, feeling that it wrecked the elegant simplicity of the equations. Whether or not he actually stated that the cosmological constant was "the biggest blunder of his life," as quoted by Russian physicist George Gamow, is debated. But he clearly regretted the move, writing, "One day, our actual knowledge of the composition of the fixed star sky, the apparent motions of the fixed stars, and the positions of spectral lines as a function of distance, will probably have come far enough for us to be able to decide empirically the question of whether or not Λ vanishes."

For many years thereafter, the cosmological constant received scant attention. The Belgian Roman Catholic Priest and astronomer Georges Lemaître referred to it in papers, as did the English astronomer Sir Arthur Eddington, but mostly the concept was forgotten. American astronomer Allan Sandage (1926–2010) even said that cosmology is "the search for two numbers."

The pendulum swung back hard to looking at the standard physics models for answers. So many people wanted to look past Einstein's universe and cling to a standard, geometrically flat, simple cosmos without these far-out relativistic changes. But the standard model had problems. In it, the critical density of the universe was perfectly balanced so that the expansion stops at time = infinity. This produces a so-called flat universe. (Although you also get gradual expansion at other densities too.) Ideas on the fate of the universe seemed to argue against a universe of simple geometry. (We will explore this concept in great detail in a later chapter.)

The Dutch astronomer Jan H. Oort (1900–1992) was busy trying to add up the contents of the universe to see if that would somehow balance the cosmos, allowing it to gently expand forever and ever, if the density were less than or equal to the critical density. It didn't seem to add up. The balanced expansion of the universe that was observed by astronomers, Oort reasoned, would require 30 times more material in the universe than astronomers observed. Including all the dark matter known in the mid-twentieth century would still leave the answer short by a factor of five. So the material required to halt the universe's expansion wasn't there, and the guess from this work was that the universe would expand forever. But something was still missing. (And it should be said that at the critical density, the expansion "halts" at time = infinity, which is effectively never.)

Oort and others then realized that the conclusion they were looking at about the open universe expanding forever and ever was dependent on *when* they were making the observations. In other words, the observations made earlier in the history of the universe than the present day might produce significantly different results. Oort was teasing around with what later came to be known as the flatness problem, the realization that some of the conditions of the early universe appeared to have very special values, and that small variations from the values would have huge effects on the later universe. It is a problem that cosmic inflation would help to deal with much later on.

American astronomer Jeremiah Ostriker and English astronomer Simon Mitton (1946–) summarized the dilemma as follows. In the mid-twentieth century, astronomers believed in the standard model as follows: "After the Big Bang the universe was, geometrically, very close to the magic flat model in which the circumference of a circle is 2π times the radius, and gravity balances the expansion to great precision. But then, much later, the universe noticed, so to speak, that gravity did not quite balance the expansion, and it became weaker and weaker until the present when it is only about one-fifth of what is needed. In the future, gravity will be negligible and galaxies will simply fly away from one another unimpeded. We simply happen to be living at just the moment of one state of the universe when gravity was quite important, to a new state in which it will be totally unimportant."

You can see how astronomers built increasingly disturbed levels of skepticism about their understanding of the standard cosmology throughout the middle part of the twentieth century. It just really didn't make sense.

In the 1970s, cosmologists began to explore options again. In 1974, a team of astronomers produced an influential paper that revived the cosmological game, pushing it beyond the then-current frustrations. It was a joint project of American astronomers J. Richard Gott (1947–), James E. Gunn (1938–), David Schramm (1945–1997), and English-born New Zealand astronomer Beatrice Tinsley (1941–1981).

Inspired by the Roman poet Lucretius (ca. 99 BC–ca. 55 BC), the group took what they believed to be a fresh look at nature and weighed the evidence for matter in the cosmos as well as the dynamics and motions. They concluded in a paper, titled "An Unbound Universe," that it is exceedingly unlikely that the actual observed density of the cosmos matched the hypothetical critical density. They also unearthed the cosmological constant, which had been ignored for several decades, and concluded that such a constant might exist. It was simply too early to tell. Tinsley later wrote a paper, before her untimely death, suggesting it was quite possible that a cosmological constant existed and could help explain a more sophisticated model of the universe than anyone had yet rendered.

Again, a lull in the machine of cosmology ensued, this time for about two decades. By the middle 1990s, the matter was again heating up. Astronomers had weighed in with better measurements of important data that go into universal models, such as the ages of the oldest stars. Still, not enough evidence could be found to support the pined-for flat model of the cosmos. But some astronomers began to explore the options that perhaps some of the universe's energy was "missing" and could account for a more comfortable cosmological model. The Hubble constant, H_0, the universe's rate of expansion, was still much argued about, and astronomers had in their minds good reasons to disagree, and not only to a trivial extent. Despite huge increases in the collection of astronomical data, and the revelations of incredible new types of objects in the cosmos over the preceding two decades, a better model of the universe was exasperatingly slow in developing.

And then came along two papers, both published in 1995. The first, by American cosmologists Lawrence M. Krauss (1954–) and Michael S. Turner (1949–), argued that timing arguments were as yet off kilter; for example, measurements of the oldest stars suggested ages greater than that of the universe! A second paper by Jeremiah Ostriker and Paul Steinhardt also pushed the same themes, relating them to the cosmological constant. If a constant exists, as both papers explored, then the universe's expansion could be accelerating over time, rather than decelerating, and much of what astronomers observed in the cosmos could then be explained in a new cosmological model. Not knowing about the cosmological constant would push astronomers into underestimating the age of the cosmos and into ignoring the fact that an accelerating universe would mean the Hubble constant would increase as time rolls on.

The cosmological constant received some revisionistic approval because of the new, fresh thinking. Many years earlier, the estimates of the Hubble constant meant that the age of the universe would be absurdly too short to make sense. The visionary Belgian astronomer-priest Lemaître realized that the cosmological constant would eliminate this problem. And the structure-growth problem of galaxies forming in the early universe would be eliminated similarly by the introduction of the cosmological constant. The team of Ostriker and Steinhardt assembled these conclusions and wrote that the reintroduction of the cosmological constant would, coupled with a universe with the critical energy density, explain the observed universe. Many other astronomers and cosmologists were skeptical.

But the tide was about to turn in a shocking observational way that would set the astronomy world abuzz. In the mid-1990s, two independent groups were busily engaged in observing very distant supernovae, exploding stars at the ends of their lives. The Supernova Cosmology Project was led by American astronomer Saul Perlmutter (1959–) and headquarted at the Lawrence Berkeley National Laboratory (LBNL) in California. This team consisted of 32 members at various institutions, including American astronomers Carl Pennypacker and Alex Filippenko (1958–) and Welsh astronomer Richard Ellis (1950–). The other team was the High-z Supernova Search Team, led by American–Australian astronomer Brian P. Schmidt (1967–) and consisted of 20 members, including American astronomers Adam Riess (1969–), Filippenko, Nicholas B. Suntzeff (1952–), and Robert Kirshner (1949–). This team was based at Mt. Stromlo Observatory in Australia, along with Riess, a postdoctoral student working with Filippenko at the University of California, Berkeley.

The two supernova hunting teams were transfixed on exploring the distant frontiers of the cosmos. When Edwin Hubble and Allan Sandage explored standard candles in galaxies in the early and mid-decades of the twentieth century, they were limited in their ability to see extraordinarily distant objects. Now, in the 1990s, the two supernova teams could use giant telescopes to push the boundaries further.

Their focus was on Type Ia supernovae, exploding stars in binary systems in which one member is a white dwarf star. The white dwarf star, a degenerate, dense, small object, will accrete mass from the companion, reaching a critical limit, the ignition temperature for carbon fusion. Within a few seconds of this nuclear fusion beginning, the white dwarf will start a runaway reaction and go off like a bomb, releasing enough energy to outshine an entire galaxy. This type of supernova releases energy to produce a relatively consistent absolute

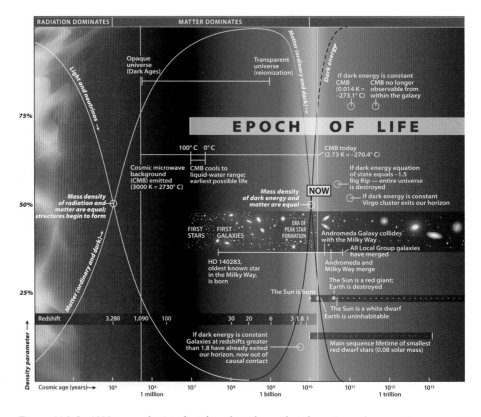

Figure 14.2 In 1998, cosmologists first found evidence that the universe's expansion was not slowly decelerating due to the force of gravity as they had expected. It is now thought that dark energy will set expansion continuing forever.
Astronomy: Roen Kelly

luminosity because the explosive white dwarfs have a uniform mass at this critical moment. This allows these objects to be used as highly accurate standard candles to measure distances to them by comparing visual magnitudes with the known absolute magnitude. Comparison of the derived distances with the redshifts then reveals the expansion history of the universe.

The two supernova groups observed a range of Type Ia objects. At the time, both groups expected to see a universal deceleration. In fact, the High-z team's credo was "measuring the global deceleration of the universe." They wanted to distinguish between eternal expansion and the possible ultimate collapse of the universe. At that time, neither team was looking for an acceleration or a cosmological constant. The results they would end up with would become a total surprise.

Both teams undertook studies at a high rate of speed. The LBNL group published their first results in 1997. They found that Ω_Λ was 0.06, which meant that the cosmological constant was vanishing. This was, at the time, the expected result.

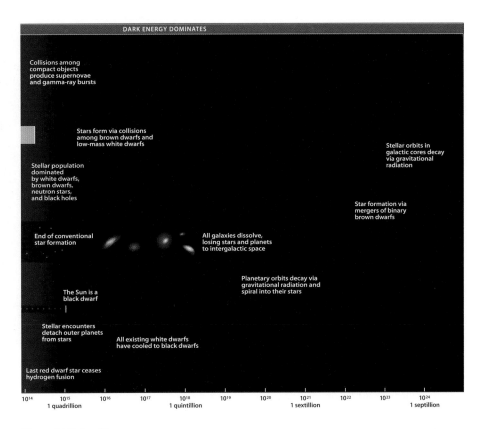

DARK ENERGY DOMINATES

Collisions among
compact objects
produce supernovae
and gamma-ray bursts

Stars form via collisions
among brown dwarfs and
low-mass white dwarfs

Stellar orbits in
galactic cores decay
via gravitational
radiation

Stellar population
dominated
by white dwarfs,
brown dwarfs,
neutron stars,
and black holes

Star formation via
mergers of binary
brown dwarfs

End of conventional
star formation

All galaxies dissolve,
losing stars and planets
to intergalactic space

Planetary orbits decay via
gravitational radiation and
spiral into their stars

The Sun is a
black dwarf

Stellar encounters
detach outer planets
from stars

All existing white dwarfs
have cooled to black dwarfs

Last red dwarf star ceases
hydrogen fusion

10^{14} 10^{15} 10^{16} 10^{17} 10^{18} 10^{19} 10^{20} 10^{21} 10^{22} 10^{23} 10^{24}

1 quadrillion   1 quintillion   1 sextillion   1 septillion

Figure 14.2 (*cont.*)

A paper by Saul Perlmutter and others would carry a great deal of weight as the research continued. "Discovery of a Supernova Explosion at Half the Age of the Universe and its Cosmological Implications," published at the end of 1997, claimed the opposite of what earlier groups had presumed, and this time based on empirical observations. This paper claimed to show, based on the group's first seven Type Ia supernovae, that the universe was decelerating and that the density parameter, Ω, was large. While this paper was soon overtaken by results in 1998, it became widely known and heavily cited and provided part of the reason that earlier, less grounded work from 1995, then became largely ignored.

The numbers of Type Ia supernovae observed along with the accuracy of the observations increased as time went on, and in 1998 the High-z team, fronted by Riess, announced a stunning result: their data demonstrated that the expansion of the universe was accelerating. The world's press immediately reacted to this sensational story, covering it extensively, and the new force in the cosmos that was causing the universe's acceleration, was formally dubbed dark energy. The following year, the LBNL team published more extensive results that also conformed to the dark energy discovery. For the work, in 2011, Perlmutter, Riess, and Schmidt shared a Nobel Prize in Physics. The leaders and the team

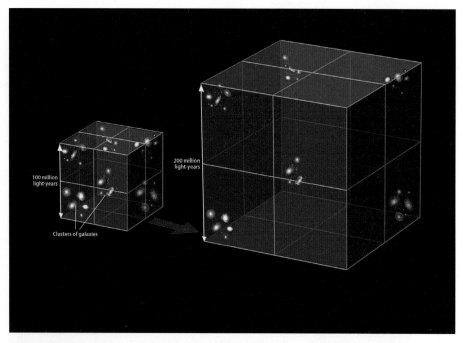

Figure 14.3 Over the past several billion years, the universe's expansion has started accelerating. Scientists are struggling to find out why.
Astronomy: Roen Kelly

members also received the 2015 Breakthrough Prize in Fundamental Physics, a prestigious $3 million award.

Both groups agreed with each other's results and painted a picture of a universe that exists with about 75 percent dark energy and 25 percent matter. The sum of it all produced what appeared to be a geometrically flat universe. Cosmologists now generally believed that we lived in a dark energy dominated cosmos. The current model of cosmology, Lambda cold dark matter (ΛCDM), was born. In the next few years, many independent facts and bits of evidence all suggested the existence of dark energy and the fact that the universe contains vast amounts of it along with dark matter. In the eyes of cosmologists, dark energy exists, and it might be the cosmological constant.

The two most recent enormously important cosmology satellites produced a huge amount of data from studying the cosmic microwave background radiation that supports and illuminates the emerging story of dark energy. The Wilkinson Microwave Anisotropy Probe (WMAP) satellite was launched into an orbit at the second Lagrangian point, some 1.5 million kilometers from Earth, in 2001. The spacecraft's objective was to study the imprint of the cosmic microwave background radiation, the echo of the Big Bang that is essentially a snapshot of the universe at 380,000 years. By carefully examining the anisotropies (fluctuations in temperature across the sky), WMAP could help to determine a great deal about the universe's geometry and evolution, as well as its overall composition.

The WMAP team was led by American astronomer Charles L. Bennett (1956–) and headquartered at Johns Hopkins University. The results from the much earlier COBE satellite pointed astronomers in the right direction (and confirmed the Big Bang itself!), but left many of the details about the universe's energy–mass composition and other details uncertain. WMAP's measurements of the microwave background radiation were instrumental in establishing the ΛCDM cosmological model, as was the growing evidence for dark energy. The data supported a universe dominated by dark energy, and in 2003 WMAP scientists announced they had the satellite's first complete year of data for public release. The WMAP data suggested an age of the cosmos at 13.7 years (when combined with data from other astronomical observations), and a preliminary composition of the universe, at the time of the CMB emission, as 10 percent neutrinos, 12 percent atoms, 15 percent photons, and 63 percent dark matter, with a negligible percentage of dark energy.

Subsequent WMAP data releases, also incorporating results from other astronomical observatories, took place in 2006, 2008, 2010, and 2012. The famous oval maps released from the mission, spotted in blue, green, yellow, and red, represented a picture of the cosmic microwave background radiation, the relic heat, released some 380,000 years after the Big Bang. The measurements of the anisotropies over the first decade of the twenty-first century provided further evidence for the existence of dark energy, and a progressively clearer picture of the universe's composition. The age was given by WMAP's 2012 release as 13.75 billion years, and the ratios of baryonic (normal) matter, dark matter, and dark energy were becoming clear.

The most precise analysis to date of the CMB commenced in 2009 with the European Space Agency's launch of the Planck satellite, named after German physicist Max Planck (1858–1947). This ambitious cosmological experiment also reached an orbit at the second Lagrangian point, and commenced studying the CMB radiation in 2010. For 3 years, Planck recorded data on the anisotropies studied by WMAP, this time improving on and complementing the earlier data. The spacecraft created a higher resolution database of the CMB fluctuations, observed gravitational lensing effects of the CMB, produced a galaxy cluster catalog using distortions in the CMB, observed a variety of active galaxies, and made observations of various properties of the Milky Way.

The increased precision of Planck over WMAP allowed improvement in mapping fluctuations in the CMB by about a factor of three – a serious step forward. Planck also produced "ovals" in false color showing a higher resolution map of the CMB relic radiation. In 2010, scientists released the first Planck data, including all-sky images and preliminary results. But the final data release in 2013 was the major one, and refined our understanding of the fundamental properties of the universe to an unprecedented degree.

The big Planck data release suggested the universe was somewhat older than previously thought, at 13.798 ± 0.037 billion years. So all of the textbooks that for years had been providing 13.7 billion years as the cosmic age needed to change their figure to 13.8 billion years. Importantly, the measurements refined the composition of the universe we now understand, such that cosmologists believe the cosmos contains 68.3 percent dark energy, 26.8 percent dark matter, and just 4.9 percent baryonic (ordinary, bright) matter. Cosmologists have a very high degree of confidence in these measurements, and they continue the high momentum of the dark energy dominated, ΛCDM model.

Throughout the discoveries about the CMB, the evolution of thoughts on the Big Bang, and the refinement of the dark energy dominant ΛCDM model of the universe, confidence continued to rise for a corollary of the Big Bang that cosmologists believe occurred during the universe's very early history. The idea for cosmic inflation arose in 1980 when American physicist Alan Guth of the Massachusetts Institute of Technology proposed the idea, as did Russian–American physicist Andrei Linde, then in the Soviet Union, completely independently of Guth.

Guth called inflation theory the "bang" in the Big Bang, which is why many cosmologists equate it with the Big Bang. In the very early universe, Guth proposed, the universe underwent an exponential expansion, by a factor about 10^{25} that happened very quickly. The patch of space in which this occurs is called a false vacuum, an unstable era of the cosmos dominated by a repulsive gravitational force. In an extremely short period, perhaps 10^{-37} seconds, the universe doubled in size, and then in just as short a time, it did so again – perhaps as many as 100 times in succession. This incredible inflation in the size of the universe occurred when the false vacuum dominated and there was virtually no normal matter. Because the false vacuum was unstable the entire period of cosmic inflation lasted just a fraction of a second before the false vacuum itself decayed, forming normal matter and dark matter.

Evidence for cosmic inflation has accumulated, in the sense that lots of evidence about the cosmos seems consistent with this corollary, and most cosmologists have high confidence in it as reality. In 2014, results from an experiment from a network of telescopes at the South Pole, called BICEP2, announced detection of the so-called B-mode polarization of the cosmic microwave background radiation, a finding that would have directly supported cosmic inflation. But soon after the press conference, other scientists began to express reservations about the validity of the findings. Nonetheless, confidence in the cosmic inflation aspect of the Big Bang remains high, and this story also connects consistently with the emerging picture of dark energy.

The overwhelming fundamental drive in cosmology is to understand dark energy, a problem numerous groups of researchers are now working on in increasingly clever ways. The dark energy density, Ω_Λ, is 0.7 and the matter density, Ω_m, is 0.3. That's the coincidence cosmologists are currently investigating.

So what exactly is dark energy? No one knows as yet, although numerous research projects have been busily working on the problem for the last 17 years. If you want to gain a Nobel Prize as a cosmologist, I suggest you get busy on figuring this out. For the present, we know that despite what appear to be some unlikely coincidences, the ΛCDM model of the universe continues to pass each test it faces when skeptics try to pull it down.

And the possibilities over the nature of dark energy are many. In 1981, the Czech–American astronomer Martin Harwit (1931–) wrote an intriguing book called *Cosmic Discovery* in which he analyzed the types of objects in the universe so far discovered and made the substantial argument that we thus far know of a relatively small fraction of the kinds of phenomena that exist in the cosmos. It is an exciting prospect to know that the types of creatures in the universe may quadruple in terms of character and complexity as astronomical science marches on, and it may be that the nature of dark energy is locked away for now in one of those many upcoming discoveries.

The very first time the expression "dark energy" went into print was in a paper by American cosmologists Dragan Huterer and Michael Turner, which they wrote in 1998 and which was published the following year in *Physical Review*. Since then, the number of research papers focused on dark energy has grown exponentially, to more than 300 per year. The problem is that we still lack observational tests or evidence to push knowledge of dark energy forward. The nature of dark energy is full of questions and lacks answers, although we firmly believe in the reliability of the concept. At present, the data are consistent with Λ, the simplest possibility. We know that a substance, energy, exists, but we don't know much about it beyond that fact.

Over the last 17 years the dark energy discovery has taught us that the universe is more complex than anyone had thought – a teaching that similarly washed over the world of astronomy and physics in the wake of Einstein's theories a century ago. The universe is not a simple place filled with matter and radiation. Instead, it is a mysterious place with this huge component of energy that is missing from our understanding of it all, and in a somewhat disturbing fashion, defined in part at least at present by this pretty straightforward concept in the cosmological constant. As the Harwit book predicted, far more of the universe is unknown than we might have believed before the dark energy discovery.

Astronomers and cosmologists do have comforting, direct evidence of key aspects of the universe – perturbations in the early inflationary cosmos, the cosmic microwave background radiation, Big Bang nucleosynthesis, the structure of particles, stars, and galaxies, the nearby cosmos we see today. But we do not yet have enough direct observations of other types in between to fully understand the frame-by-frame nature of the history of the entire universe, or its composition – in large part, the nature of dark energy and how it has interacted with the cosmos over the last 13.8 billion years.

Like a group of revelers who have awakened after a big weekend bash, astronomers and cosmologists have been shocked in part about how little we

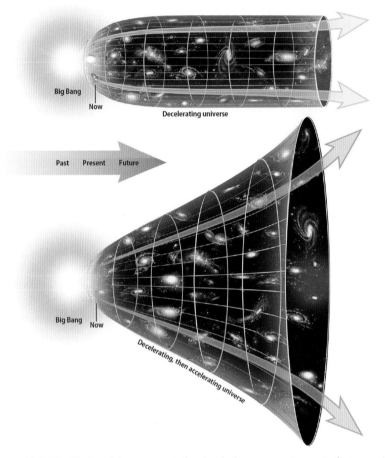

Figure 14.4 The likeliest future scenario for the dark energy universe is that we will continue to expand forever (bottom), leading to eventual heat death and a so-called Big Freeze.
Astronomy: Roen Kelly

know compared to what we thought we knew. There is a bit of a confidence hangover that has allowed some new and unusual avenues of research to push forward in the quest to define the nature of dark energy.

Avenues from particles physics have kept theoreticians extraordinarily busy. Could string theory, in which the ordinary particles in subatomic physics are replaced by one-dimensional objects called strings, be part of the answer? Could quintessence, a hypothetical form of dark energy that has been pushed by some as a fundamental fifth force, explain some of what we see? Could dark energy be some form of a modification of gravity? Could we see evidence of cosmic acceleration in the universe without the need for dark energy? A whole host of possibilities has cosmologists and particle physicists scrambling in many directions, some of which could turn out to be extremely helpful in the new quest to understand the complexity of the universe.

The drive to define dark energy has spawned a whole host of dedicated dark energy research projects. Just one of the many, for example, is the Dark Energy Survey, a multinational effort that is employing the 4-m Blanco Telescope at Cerro Tololo Inter-American Observatory in Chile to image large areas of the southern sky, investigating large-scale structure by examining Type Ia supernovae, baryon acoustic oscillations (regular, periodic fluctuations in the density of the visible baryonic matter of the universe), galaxy cluster counts, and weak gravitational lensing (the bending of light from distant objects by foreground galaxies). Astronomers involved with the Sloan Digital Sky Survey are similarly investigating baryon acoustic observations. And many other groups are involved in such cosmological investigations, which will no doubt continue to narrow down the standard cosmological parameters, and in doing so, we hope, make a breakthrough discovery about just what dark energy is.

These efforts will also push the techniques of understanding the distant universe to higher levels, so that astronomers, in the quest for dark energy, will also refine their abilities to measure distances in the cosmos, explore the limits of the most distant objects we can see, increase the precision of our understanding of how the universe works, and allow a refined picture of how we interpret astronomical data from many other sources. All of this will make astronomy a stronger science in the future. And if we're lucky, all of it will build a stronger picture about the early history of the cosmos, and perhaps about the meaning of it all – about something as fundamental as why we're here.

Chapter 15
Black holes are ubiquitous

The best summary line about black holes I've ever heard came from the American theoretical physicist Kip S. Thorne (1940–), who described them during his talk at the first Starmus Festival in 2011: "The brightest objects in the universe – but no light!" That simple statement reveals much about the bizarre nature of black holes, objects of such intense gravity that nothing – not even light – can escape from them. Black holes are certainly among the favorite and most alluring objects in the universe for many astronomy enthusiasts, the subject of almost endless mystery and speculation. The reality of black holes is even stranger than the limited ideas many people have about them.

The knowledge base about black holes is of very recent vintage. When I joined the staff of *Astronomy* magazine in 1982, black holes were mostly a rumor. (The Milky Way black hole candidate Cygnus X-1 had been identified but not confirmed as a black hole; and astronomers believed quasars contained black holes, but had not yet confirmed that.) That was the case despite the fact that the conceptual ideas for these objects stretch back more than 230 years, to the English natural philosopher and clergyman John Michell (1724–1793). In a 1783 paper he wrote for the *Philosophical Transactions* of the Royal Society in London, Michell proposed the idea of "dark stars," objects of such strong gravitational pull that nothing, including light, could escape them, and therefore the stars would be invisible.

In a visionary way, Michell even wrote that because such stars would be invisible, they would have to be detected from their effects on other objects around them, as with stars in a binary system in which one star had become "dark." Several years after Michell's ideas were published, the great French mathematician and astronomer Pierre-Simon Laplace (1749–1827) also wrote about the same concepts in a book published in 1796.

Yet the evidence for black holes was extraordinarily slow in arriving. And the term black hole was not yet coined until relatively recent times. In fact, there was essentially no progress in the hazy concepts outlined by Michell and Laplace for more than a century, and it was only our old friend Albert Einstein who ignited substantial interest in the concepts again with his general relativity

Figure 15.1 Our galaxy is home to a supermassive black hole at its center inside the compact radio source called Sagittarius A*, but unlike some monstrous black holes seen in other galaxies, ours is not a voracious eater.
NASA/CXC/MIT/F. K. Baganoff, *et al.*

and with its notions that space-time was warped rather than nice and flat. Black holes were certainly predicted from Einstein's theories, although they were at the time called "Schwarzschild singularities" after Einstein's friend, theoretical physicist Karl Schwarzschild, who had worked on the problem. In 1939, Einstein wrote a famous paper in which he discussed the prediction of these objects and also fretted that they had not been observed, and contemplated why "they do not exist in physical reality." Einstein had rejected his own idea a little too soon.

Schwarzschild, although he did not live long, in his calculations laid down the basis for much of what we still believe about black holes, which came to be known as Schwarzschild geometry. Working from Einstein's general relativity, Schwarzschild created a geometric concept that included the important notions of relativity and also of space being a mixture of space and time – which followed Einstein. Objects in the universe warp space around them but they also warp time, and so space and time are seen differently from the viewpoints of different observers at different places. To help sort this out, physicists use so-called embedding diagrams to visualize the interplay between objects, observers, space, and time, and Schwarzschild's geometry plays a key role in this process.

Imagine a star with a flat, 2-D rubber sheet stuck through its equator. Now, let's imagine that we replace the third spatial dimension – the one that is perpendicular to the sheet – with something different, with a measure of how strong the star's gravitational pull is WITHIN the sheet, expressed geometrically (as Einstein thought of gravity) as a vertically downward stretching of the 2-D sheet. In our pictorial analogy, the sheet (held in our hands, here on Earth) is stretched as though we laid a heavy ball on it. Einstein understood the vertically stretched rubber surface as a representation of how gravity curves that space in 3-D. THAT curvature is not easily imagined. We use our rubber sheet to help us imagine it.

Schwarzschild's great contribution with his equations was to describe not only the warping of space around objects with mass in the universe, but also the way in which time is warped around them, due to the star's gravity. The warping of time near a star is described by time dilation, a concept originally introduced by Einstein. Near a star's surface, time passes more slowly than at a point far away from it.

After Schwarzschild's early death from disease acquired at the Russian front during World War I, physicists, including Einstein, embraced his equations, and his geometry of space-time began to be a central part of physics. It became consensus that large stars like the Sun would produce a curvature of space-time around them, and that light emitted from the stars would be very slightly shifted in color toward the red. The more compact and denser the star, the greater would be the curvature of space-time, all agreed. And pushing this compactness to the extreme would reawaken the notions of "dark stars" from their dormancy since the days of Michell and Laplace.

Michell and Laplace lived in a world of Newtonian mechanics, not yet unbridled by Einstein. They believed that light particles nearly escaped very small, very dense stars, but were pulled back by gravity and therefore didn't escape outward into the cosmos. As Einstein had proposed, Schwarzschild believed that light always moves at the same velocity and cannot be slowed or reversed in direction as his forebears thought. He showed that light must be shifted in wavelength by an infinite degree as it travels upward an incredibly small distance from a highly dense body. The infinite shift in wavelength is what cancels the light from "existing," from being seen. All of the light's energy is gone. No light whatsoever is emitted from such a star.

During the generation that followed Schwarzschild's death, however, the two most influential relativists, Albert Einstein and Sir Arthur Eddington, did not take Schwarzschild's ideas seriously. The concept of black holes was simply too strange. They seemed to be too weird even for relativistic physics. The intuitions of the world of physics, led by Einstein and Eddington, simply would not admit such revolutionary notions.

During the 1930s, the momentum of belief in black holes accelerated among physicists, but Einstein himself published a paper in 1939 that concluded black holes could not exist. His calculations were well done, but his interpretation of them was not. More advances in understanding the deaths of stars would have to take place before the belief in black holes caught on more widely.

Influenced by Eddington, Indian astrophysicist Subrahmanyan Chandrasekhar (1910–1995) began to study white dwarf stars, the highly dense remnants of stars like the Sun, as a student at the University of Cambridge in the early 1930s. Chandrasekhar in fact confronted the problem of how white dwarfs shrink as degenerate stars, something that Eddington had studied and could not solve, as he believed their density was 10,000 times less than it really is.

Chandrasekhar found a solution by ignoring the laws of physics that Eddington was bound to and thinking in terms of the white dwarf star's demise in quantum mechanical terms. This described how the matter inside a white dwarf was squeezed into incredibly high densities. So-called degenerate motion of matter inside the star, achieved at 10,000 times higher densities than rock, follows from the wave/particle duality of particles under these conditions.

Chandrasekhar also demonstrated that the mass of a white dwarf star could not exceed 1.44 times the mass of the Sun, thereby establishing the so-called Chandrasekhar limit. The limit is the mass above which electron degeneracy pressure in the star's core cannot balance the star's gravity, now refined to 1.4 solar masses. In 1935, Eddington publicly derided Chandrasekhar over this point, but it was the young Indian astrophysicist and not the senior English astronomer who turned out to be right.

About the same time as the work on white dwarf stars was pushing forward, Swiss astronomer Fritz Zwicky, who we have met before, was busily working on the physics of supernova explosions. Along with German astronomer Walter Baade (1893–1960), Zwicky set up shop with Schmidt cameras to image exploding stars – and they coined the term supernova for them. The almost concurrent discovery of neutrons seemed made to order: Zwicky believed that the degenerate core of a star might implode until it reached a density approximately equal to an atomic nucleus, and the material in the shrunken stellar core would resemble a gas of neutrons. And he called the concept a neutron star.

Zwicky calculated that the star's core would be so tightly bound by gravity that it would be incredibly small and its mass would also be reduced, by 10 percent. That 10 percent of the star's mass, Zwicky reasoned, would go into explosive energy, and the degenerate star would go off like a bomb – a supernova. The star's outer layers would fly off at high speed and leave a degenerate stellar remnant of some type behind. He also took note of cosmic rays and believed the total energy detected from them was roughly equivalent to the energy released by supernovae – and time would prove him correct. Exploding stars were the source of distant cosmic rays. The processes that lead to supernova explosions are several and varied, beyond the simple conversion of gravitational energy into explosive energy. Many details of nuclear physics make the supernova bang. Superheating of material during the star's collapse, for example, also powers the explosion.

Gradually, through Einstein, Eddington, Chandrasekhar, Zwicky, and many others, a picture of how stars die began to come together, solidifying in large part by the second half of the twentieth century. The Chandrasekhar limit plays a critical role. As we have seen, for a star like the Sun, a remnant white dwarf star is the ultimate fate. As a sunlike star dies, gravity begins to win out, pulling the star into a compressed mass, squeezing the electrons into a smaller and smaller space until it can compress no more, as the electrons fight back with degenerate pressure. The white dwarf stellar graveyard is established.

For stars more massive than the Chandrasekhar limit, however, things become more interesting. As a star like this cools and shrinks, its electrons are likewise squeezed into smaller spaces, and the degenerate pressure rises, but the electrons are more powerless in a more massive star. In this case, the star's gravity continues pushing the matter inward, which then becomes a neutron star, and the internal pressure is produced by neutrons. Again, mass plays a key role in the outcome here. For many stars, they will simply settle as neutron stars, gradually cooling as they achieve an uncomfortable stability of sorts.

Astronomers know there is a range of masses in which stars die as neutron stars, and a range of greater masses in which stars die as black holes. There is probably a maximum mass that any neutron star can have, and beyond that mass, gravity would overwhelm the neutron pressure. For masses far greater than the greatest neutron star mass, dying stars will form a black hole. The density of a neutron star is about 10 quadrillion times that of water – a mass perhaps comparable to that of the Sun, but in a diameter of only 20 kilometers across.

So the evidence points toward the existence of stellar black holes, those derived from the deaths of stars, as resulting from stars that were 10 times or greater than the mass of the Sun. The black hole itself is a region of space-time in which gravity prevents everything, including light, from escaping. And the evidence suggests that a stellar black hole may be of the order of 30 kilometers or so across. They are very tiny!

For some massive stars, it's possible that a collapse leads directly to the formation of a black hole, without an explosion. Some astronomers believe that such stars may "wink out" and a recent report suggests that an event like this has been detected. But this idea is controversial and unconfirmed. For moderately massive stars, in the range of ten to 100 solar masses, the situation is more complex. These stars form iron cores that absorb energy and trigger the star's collapse. The neutron star forms first, and the energy released in that process turns the implosion into an explosion. How this reversal happens is not yet exactly understood. If some of the material from the explosion falls back onto the neutron star, this may spark the formation of a black hole.

The theoretical work to decipher black holes continued apace. No one knew they really existed until very recently. A critical year occurred in the development of black hole ideas in 1958. In that year, American physicist David Finkelstein (1929–) identified the hypothesized surface of the black hole Schwarzschild had imagined as an event horizon, a "perfect unidirectional membrane – casual influences can cross it only in one direction." Envisioning what black holes might be like took a step forward. Also in that year, American physicists John Archibald Wheeler (1911–2008) and J. Robert Oppenheimer (1904–1967), both veterans of the Manhattan Project, battled over the ways in which degenerate stars imploded. Their arguments led to a better understanding of the physics of exploding stars and the particle physics behind black hole birth. It was Wheeler, in fact, who invented the name black hole.

The 1970s saw an explosion of progress in thinking about black holes. A decade earlier, little was known about them. By 1975, Chandrasekhar was pioneering another revolution, assembling models of black holes based on three main properties: mass, electrical charge, and spin. The great physicist, who had won the argument with Eddington, now understood that with an understanding of these properties, he could calculate the strength of a black hole's gravitational attraction, the shape of the black hole's event horizon, and the degree of distortion of space-time surrounding it.

Three great physicists led the day during this period, and also influenced a legion of younger followers who would carry on black hole research. They were Wheeler, Russian Yakov B. Zel'dovich (1914–1987), and Englishman Dennis Sciama (1926–1999). Among the discoveries of this period was that black holes "have no hair," a phrase coined by Wheeler, meaning that nothing "sticks out" of the black hole from which clues might be offered about the progenitor star or anything else. However, black holes can spin, and it was New Zealand physicist Roy Kerr (1934–) who detailed equations that demonstrate how black holes spin. Kerr showed that a spinning black hole creates a tornado-like swirl in the space-time that surrounds it. There is a physical difference between non-spinning black holes that are spherical and spinning (sometimes called Kerr) black holes that have an elongated ergosphere, the region outside a rotating black hole.

Strange things happen when a spinning star creates a spinning black hole. A particle falling into the spinning black hole follows the swirl of space around the black hole but then falls into a locked position, rotating with the event horizon itself. As seen from the outside, the particle travels around and around the horizon forever. From the particle's point of view, its clock moves at the normal rate. But to an outside observer, watching the particle, time is dilated and the particle appears to be moving slowly. In the "usual time" from the particle's point of view, the particle sinks into the horizon and toward the black hole's center.

Black holes also can pulsate. One of Thorne's students, William H. Press (1948–), proposed in 1971 that the ripples of space-time around a pulsating black hole can be conceptualized as pulsations of the black hole itself. Press also found that black holes pulsate at certain frequencies, just like stars, through employing computer simulations. He thus identified the pulsations of black holes and made the analogy to the ringing of a bell.

Up to this point, nearly all of the work on black holes, from the first concepts by Michell in 1783 onward, had been conceptual – theoretical. This made enormous numbers of folks skeptical, of course, despite some of the finest minds in the history of science being really sure that black holes exist in the universe. But where were they?

Of course, astronomers have always faced significant trouble with detecting black holes because they are black – you can't see them. The theoretician who showed interest in getting astronomers to search for black holes was Zel'dovich. But how to begin? Black holes are not only black, but they are physically tiny – tens of kilometers across. At a distance of 4 light-years, the nearest star to the Sun, a black hole would be imperceptible. Lensing of a distant object by a black hole would also be a difficult way to detect one. But Zel'dovich reasoned that if a black hole were in orbit with a companion star, the star could betray the black hole's presence by Doppler shifting its light alternately blue and red as it orbits the black hole. Looking for a star with telltale spectral line shifts, coupled with a calculated mass, could reveal a black hole. Zel'dovich also realized that black holes might reheat some of the surrounding infalling matter, and therefore would be bright x-ray emitters. So the search was on for binary systems and also stellar objects that were x-ray bright.

X-rays do not penetrate Earth's atmosphere, and so eventually astronomers launched satellites to observe the x-ray sky. Two of them, *Uhuru* in 1970 and *Einstein* in 1978, created some interesting observations. One of the most alluring was a star system dubbed Cygnus X-1, which seemed the most promising black hole candidate. Thorne and his friend English theoretical physicist Stephen Hawking (1942–) made a bet, Thorne wagering that Cygnus X-1 would turn out to be a black hole. By 1990, Hawking conceded the bet, as the evidence seemed extremely strong that Cygnus X-1 was indeed a stellar-mass black hole.

Cygnus X-1 was just what the doctor ordered in those very first search concepts. It was a binary star located 6,000 light-years away and is composed of a bright (9th magnitude in our sky) O-type star along with an x-ray bright, optically dark companion. The dark companion "weighs" about 15 solar masses. It can be thought of as the first black hole discovered – although reasonable confirmation that Cygnus X-1 is a black hole took at least 15 years following the object's discovery. And the certainty that Cygnus X-1 is a black hole is extremely high, perhaps 95 or 98 percent, but astronomers are not yet absolutely certain because other exotic explanations for a massive, dense object in its place could exist. But the overwhelming odds are on the black hole explanation.

Figure 15.2 Extreme 1 million-degree temperatures in Cygnus X-1's accretion disk make it a strong x-ray emitter; it was the first strong candidate for a stellar mass black hole. NASA/CXC/SAO

The nature of the evidence required a team of hundreds of physicists, astronomers, and theoreticians to develop evidence for the black hole's existence throughout the 1960s and 1970s. The picture gradually developed of a blue supergiant star, HD 226868, with a mass of about 30 times that of the Sun and a diameter some 20 times larger than our star. The supergiant and the black hole orbit a common center of mass once every 5.6 days, and the distance from the star's surface to the black hole is about 14 million kilometers.

Whereas neutron stars can send an unambiguous signal about their identity, a pulsed beam of radiation that could only be explained by their nature, the fact that black holes don't have "hair" keeps them a little mysterious. It may be that black holes will always be detected by inference and not directly by anything like associated radiation. But it's possible that unusual situations, like pulses of radiation from a gas cloud orbiting a black hole, could give away ironclad, "direct" evidence. But even without the direct evidence, the inferred evidence about black holes leaves astronomers virtually certain of their existence, and well beyond that of simply Cygnus X-1. The Milky Way Galaxy holds other black holes in similar binary systems, as astronomers now know.

Direct evidence could come from elsewhere, too – conceivably, a rotating black hole could radiate from radii inside the Schwarzschild radius of a non-rotating black hole. Seeing radiation from this type of source would be one bombproof way of directly testing whether the objects in question are black holes.

Also in the era of the 1960s and 1970s, astronomers began to investigate possibilities that black holes are not made by stars alone, nor do they exist simply on the scales of stars. Even before this time, in 1932, American physicist Karl Jansky (1905–1950) pioneered radio astronomy with a large antenna he crafted, detecting a radio hiss emanating from what seemed to be the center of the Milky Way Galaxy. Astronomers were perplexed. They expected the Sun to be the brightest radio source and could not explain how the center of our galaxy could be. Many years passed as astronomers slowly gained interest in this problem and as instruments gradually improved.

Finally, in the early 1960s, radio astronomy seemed to be pushing forward in a more serious manner. Various radio surveys turned up strangely bright objects scattered across the sky, and these surveys were principally conducted by English astronomer Martin Ryle (1918–1984) at the University of Cambridge. In 1960, this group discovered a peculiar object called 3C 48 (the 48th object in the Third Cambridge radio catalog), and its spectrum was unlike anything they had ever seen. Moreover, it looked like a tiny blue point of light, and not a diffuse radio galaxy. In 1963, however, this mystery resolved when Dutch astronomer Maarten Schmidt (1929–), working at the California Institute of Technology, had a moment of realization. He had a similar strange spectrum of 3C 273, and he suddenly realized that telltale spectral lines were shifting by 16 percent – this object must have contained plentiful hydrogen and be moving away from Earth at the breathtaking speed of 16 percent of the speed of light.

Schmidt had identified what astronomers called quasi-stellar objects, or quasars. And the meaning of quasars was staggering: Schmidt realized he had identified the most luminous object in the universe, an object that was 100 times brighter than the brightest galaxy ever seen, and that it was at the amazing distance of 2 billion light-years, and was moving incredibly fast. It took some years to nail down that the energy source was very tiny. Astronomers eventually realized that quasars were producing an incredible amount of light in a very small volume of space – perhaps a million times smaller than the Milky Way's disk. What could possibly explain this incredible combination of extremes?

The answer tied back to Jansky's discovery of radio noise from the center of the Milky Way. A number of astronomers had proposed that this energy could be explained by synchrotron radiation, produced from electrons being accelerated to near the speed of light and being ejected outward along magnetic field lines surrounding the galactic center. As they were grappling with this issue, astronomers continued to discover more radio galaxies, and many of them had double lobes on either side of the galaxy's center. These, too, seemed like candidates for synchrotron radiation. The energy involved with these objects was so amazing, though, that it left astronomers looking for answers in the extreme realm of astrophysics.

By the 1980s, astronomers were imaging radio galaxies with far higher resolution, using a technique called Very Long Baseline Interferometry, which allowed tying together multiple radio telescopes to increase resolution. This enabled them to see the jets shooting outward in radio galaxies, that formed radio-bright lobes, were sharply focused, and extended right down into the centers of the galaxies. It became clear that a "central engine" of very great force had to be powering these jets, and of the many proposed explanations, only one passed all of the logical tests to explain what was going on – supermassive black holes living in the centers of these galaxies.

The idea originated in the early 1960s and was perfected in 1969 by English astrophysicist Donald Lynden-Bell (1935–), who suggested that gas falling inward toward a black hole spirals again and again, forming a so-called accretion disk before some of the gas passes into the black hole. The accretion disk spirals rapidly and violently, and the gas within it heats it incredibly. Astronomers realized that 3C 273's bright, tiny center was caused by radiation from the accretion disk.

Moreover, there are four possible ways in which this supermassive black hole could produce the double-jet seen in numerous high-energy galaxies. The black hole could be surrounded by a cloud of cool gas, through which a smaller bubble of hot gas could punch holes, emitting a focused jet. Or the high internal pressure of the disk could produce whirlpool-like funnels in the disk. Or magnetic field lines anchored in the disk could focus gas outward and upward in jets. Or the black hole could be threaded by magnetic field lines, dragging them as it spins, causing the gas to be whipped around rotationally, shooting matter outward. Jets are observed to be up to several million light-years long, and many of them are straight over their whole length. This means the central engine remembers an ejection direction for several million years. Rather than energy punching randomly through surrounding gas, black hole jets are more

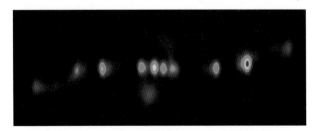

Figure 15.3 The peculiar object SS 433 is home to an ordinary star orbiting an extremely compact object, which astronomers suspect is a neutron star or a black hole. The object is surrounded by an accretion disk that spews radio-emitting jets.
Amy Mioduszewski *et al.* (NRAO/AUI/NSF)

like gyroscopes, with aim that is hard to tilt and a driving engine that involves spin and magnetic fields.

As modelers worked on the dynamics of how black hole jets and accretion disks could work, astronomers continued to discover more and more so-called active galaxies, or active galactic nuclei (AGN). Throughout the 1980s and 1990s many dozens of AGN turned up in surveys and individual observations as equipment and techniques improved.

In 1988, astronomers published the first convincing evidence for dynamically identified black holes more or less simultaneously with two studies of the Andromeda Galaxy, one by American astronomer John Kormendy (1948–) and another by American astronomers Alan Dressler (1948–) and Douglas Richstone, both studies using ground-based observations. In the same year, soon thereafter, the existence of two more black holes was confirmed from ground-based observations, in the Sombrero Galaxy (M104) (again by Kormendy), and in the galaxy NGC 3115 (by Kormendy and Richstone).

These three, and two more black holes found by Kormendy, all with the 3.6-m Canada-France-Hawaii Telescope on Mauna Kea, Hawaii, were all later confirmed by ground-based work and with the Hubble Space Telescope. So the first five black holes confirmed resulted from ground-based telescopes. And so did the most convincing cases of supermassive black holes, which included those in the Milky Way and in M106.

Astronomers using the Hubble Space Telescope made many observations of active galaxies in the 1990s, and advanced black hole astronomy from a "proof of concept" stage to a revolutionary time in which astronomers could study plentiful supermassive black holes in reasonably large numbers. The crucial work of relating black hole mass with the luminosity and mass of the host galaxy's central bulge took place on the ground. By the late 1990s, work suggested strongly that central black holes exist within nearly all normal galaxies – except for dwarf galaxies – and that most black holes, in the recent universe, are quiescent, or sleeping. There simply isn't enough matter falling into them any longer to awaken them into a fury of energetic outflow.

But earlier in the universe, most galaxies were highly active with supermassive black holes feeding on infalling gas, stars, and dust, and with a violent burst of energy flung outward from material spun around the accretion disk that did not fall into the hole. This produced quasars, the energetic centers of young galaxies, and a whole spectrum of AGN that were thought for years to be different types of galaxies with different phenomena, but are most likely the same black hole phenomenon seen from different angles and frames of reference from our point of view.

In recent times, 2013 to be exact, several astronomers summarized the current state of research on supermassive black holes. A particularly great review came from Kormendy and colleague Luis C. Ho (1967–). They described how supermassive black holes have now been found in 85 galaxies by dynamical modeling of the kinematics of material surrounding them. They summarized how

astronomers have found a tight correlation between the mass of the supermassive black hole and the velocity dispersion of the host galaxy's central bulge.
Astronomers think that central black holes in galaxies and the galaxies' bulges
evolve together by regulating each other's growth.

But as more research has taken place, the story has become somewhat more
sophisticated yet. In actuality, black holes correlate differently with different
types of galaxies. As is often the case, nature turns out to be richer and more
complex than was originally suspected. Black holes with 100,000 to 1 million
solar masses exist in many galaxies without central bulges. And galaxy disks
are not necessarily correlated with black hole formation.

Kormendy and Ho found four major types of supermassive black hole
evolution in galaxies. First, in largely bulgeless galaxies, small black holes can
form and grow by local, secular, episodic, and stochastic feeding, and this
involves too little energy to allow the galaxy and black hole to evolve in some
paired relationship. Second, global feeding in galaxy mergers can produce
enormous, quasar-like black hole energy output that does help to control the
galaxy's evolution. This process makes classical bulge galaxies and disky elliptical galaxies. Third, giant elliptical galaxies have huge black holes that occasionally light up as mini-quasars. Their energy output keeps lots of matter
locked up in gas. These objects were probably formed by an earlier generation
of galaxy mergers. And, fourth, after many successive mergers, in each of which
the galaxies add together and their black holes add together, it is inevitable that
galaxies and their black holes will look closely related.

Of the many supermassive black holes discovered in galaxies since the 1990s,
one of the most celebrated is the one in M87, the giant elliptical galaxy located at

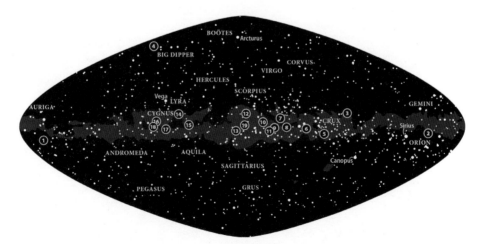

Figure 15.4 Nineteen black holes are shown here in our galaxy, but that number is expected to
grow significantly over time. Most of these black holes lurk along the densely populated
galactic plane.
Astronomy: Roen Kelly

the core of the Virgo Cluster. Brightly visible as a fuzzball in backyard telescopes, this remarkably bright galaxy has a well-defined jet emanating from its central black hole that is focused enough to be imaged with large amateur telescopes. Lying at a distance of 53 million light-years, the galaxy holds some 200 times the mass of the Milky Way and spans 500,000 light-years across. The jet itself is some 5,000 light-years long, and the galaxy's black hole holds some 3.5 billion solar masses, nearly 2,000 times more massive than the black hole at the center of the Milky Way.

The M87 black hole jet, like many others, shows well-defined knots that appear to move faster than the speed of light. In the case of M87, they appear to be moving about 6 times faster than the speed of light. English astronomer Martin Rees demonstrated what this means: when the jet is aimed almost exactly toward us, the velocities of almost the speed of light look like velocities greater than the speed of light. This is important, because it is part of the proof that central engines are black holes. Only a black hole is compact enough to fling out material at 98 or 99 percent of the speed of light.

While the inventory of black holes in the universe increases over time, so do the strange concepts attached to them. In the early 1970s, Stephen Hawking demonstrated in full, following work and discussions with Mexican–Israeli theoretician Jacob Bekenstein (1947–) and Yakov Zel'dovich, that black holes can radiate. According to the quantum mechanical uncertainty principle, black

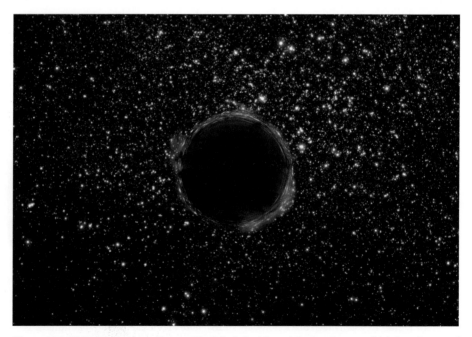

Figure 15.5 This artist's view shows a black hole inside a globular cluster, which is only one of the many unusual places astronomers have uncovered these gravitational monsters.
NASA and G. Bacon (STScI)

holes can lose mass by creating and losing particles, meaning they would slowly evaporate over time. This would be particularly true of the hypothetical micro black holes, tiny black holes that compress the mass of the Moon into a volume too small to see. Thus, this so-called Hawking radiation, possible from very low-mass black holes, means that at least some black holes could not live forever – that even they would disappear over time. And if the universe lasts forever, even stellar-mass and supermassive black holes will eventually evaporate.

To turn Earth into a black hole, by the way, you would have to squeeze it down into the size of a grape.

And what of the most popular science fiction aspect of black hole theory, that so-called wormholes, warps in space related to black holes, could allow aspiring astronauts to travel suddenly to other places in the universe or to travel forward or backward in time? This most intriguing question related to black holes, the staple of sci-fi and tabloid blogs, may not be as alluring as we all would like to think. Black holes are a natural consequence of stellar evolution. Wormholes are a remote hypothetical possibility of Einstein's field equations and would require two singularities of just the right type to form at close proximity, an unlikely event, and relativity theory suggests that any wormhole would exist for a very short duration if at all. Moreover, there is no reason to believe that the universe today contains any singularities of the kind that would breed wormholes at all – no evidence whatsoever.

Entering a wormhole or a black hole and surviving is also a nonstarter in reality – the black hole, at least on stellar scales, would pull you into a string of protons 10 kilometers long, so who cares where you might come out? It would pretty much ruin your day, regardless of the possibilities of travel in time or in space. But the smart money is on Stephen Hawking's prediction: "Whenever one tries to make a time machine, and no matter what kind of device one uses in one's attempt (a wormhole, a spinning cylinder, a "cosmic string," or whatever), just before one's device becomes a time machine, a beam of vacuum fluctuations will circulate through the device and destroy it." Among cosmologists, the use of wormholes or black holes to time travel or to travel efficiently to other places in the universe is decidedly out of fashion. It is just not something that is realistic at all, despite the intriguing theoretical possibilities raised by Einstein's field equations.

So black holes leave us with being the brightest objects in the universe that we can't see – they emit no light. The groans of immense outpourings of gravitational waves, sent rippling through space-time when black holes collide, slosh throughout the universe as loudly as they can. And yet with our earthly eyes, focused on the visible part of the spectrum, we do not see them. But over the past generation we have come to know they are there, for certain, and to know that they intrigue us like almost nothing else does.

Chapter 16
What is the universe's fate?

Astronomy and astrophysics are filled with countless questions, and are slowly gaining some pretty impressive answers. Certainly, one of the most fundamental questions of all, one that stretches back perhaps the longest in philosophical terms in the human mind, is one of the simplest: What will become of the universe?

This elegant question is not an easy one to answer. We know that as we look out into space, we're looking back into time. The distant universe is a snapshot of what existed billions of years ago, and we do not have an accurate picture of many of the objects we see as they really are now, at this exact point in time. Knowing the status of objects in the universe in the "here and now" works very well for our solar system, for Earth, the Sun, and our family of planets, asteroids, and comets. But as we look progressively out even into our Milky Way Galaxy, we begin to see things as they were, more so as distances increase. So how do we predict what will happen well down the road in the universe's future? To predict how the cosmos will end? It is a stupefyingly difficult problem.

To predict the universe's future, astronomers would like to know all about its current physical parameters, as well as the models of cosmology that they believe most strongly in. We have already seen that the current size of the universe is at least 46 billion light-years in each direction from where we are, and that it could be larger yet if cosmic inflation theory is correct, which nearly everyone believes is so. What about the shape of space? According to general relativity, space is curved by a degree set by the average density of matter within it. So exactly how dense the universe is becomes a really big question. Space might be flat, positively curved like a balloon, or negatively curved like an equestrian saddle.

But actually measuring the shape of space is really tough. Measuring distances to and velocities of countless stars and galaxies failed to provide the answer to the universe's shape. Instead, cosmologists now look to the cosmic microwave background radiation, and subtle variabilities within it, to infer the shape of the cosmos. As we have seen, the radiation is an imprint of the

Figure 16.1 The central galaxy of the Phoenix Cluster, positioned about 5.7 billion light-years from Earth, is one of the largest known objects in the universe. The object puzzles astronomers because star formation has slowed to a crawl in most galaxies as the universe evolves. NASA/CXC/M.Weiss

universe at around 380,000 years after the Big Bang, and so the temperature fluctuations measured by WMAP and Planck within the afterglow provide essential clues.

The clues are there only because the cosmos can act as a gigantic lens. Temperature variations in the cosmic microwave background would be magnified variably depending on the curvature of space. Recent measurements show that the curvature of space is very slight, less than 1 percent, and so the universe is nearly flat. But that's not the whole story. Globally, the universe might not be close to flat, and we don't yet have any ways to test that idea.

The murky story of what we know about the contents of space – baryonic (normal) matter, dark energy, and dark matter – relates to the future of the universe. In the early cosmos, the universe was awash in photons and other particles moving in a uniform bath at nearly the speed of light. At a point some 55,000 years after the Big Bang, the densities of energy and matter were balanced. As the universe aged, however, the density of matter decreased but the expansion of the cosmos caused the density of energy to drop more quickly. Whereas radiation dominated the early universe, it is hardly the dominant player now.

By the time of the cosmic microwave background radiation release, at 380,000 years after the Big Bang, the universe was awash in an overall temperature of about 2700 °C (4900 °F). Some 200 million or more years later stars and

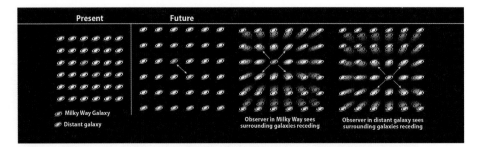

Figure 16.2 The Big Bang set the universe in motion and it has been expanding ever since. In recent decades, cosmologists have found increasing evidence that galaxies are speeding ever farther apart.

Astronomy: Roen Kelly

galaxies started forming, but no one yet knows the sequence – did stars form and then fall into galaxies by gravity, or did galaxies assemble, allowing stars to wink on, or did black hole "seeds" form first, and then the galaxies around them? That is a question the planned James Webb Space Telescope may be able to answer.

Strangely, the great era of star formation is mostly now over. Galaxies came together and merged, time after time, and set off great rounds of star formation within the largest galaxies, peaking some 3 to 4 billion years after the Big Bang. Star formation in smaller galaxies occurred more slowly and generally later, with a high point some 7 or 8 billion years after the Big Bang. In our current time, 13.8 billion years after the bang, the star forming that goes on is relatively minor.

Because the universe is expanding so rapidly, the formation of new structures in the cosmos is generally getting to be harder and harder as time rolls on. More distance lies between most clumps of normal matter – galaxies – every day, and much of the gas that would go into making new stars is increasingly left in the middle of nowhere, in interstellar or intergalactic space, where it can't get together with lots more stuff to build more structures. And, of course, the stunning discovery of the last 17 years is that dark energy is accelerating the expansion of the cosmos as time goes on. We must remind ourselves that in our quest to know about the fate of the universe, the matter we know the best, normal matter, is just 5 percent of everything, despite the fact that it makes all the stars and all the 100 billion galaxies we see. The real mysteries, dark matter and dark energy, hold the keys to the cosmic future.

The discovery of dark energy in 1998 by the two broad teams of astrophysicists, and led by Saul Perlmutter, Adam Riess, Brian Schmidt, Alex Filippenko, and others, revolutionized understanding of the end of the cosmos. The fact that the universe is expanding at an accelerating rate is deviously hard to explain, and the cosmologists who figure it out would certainly be in line for a Nobel Prize, among other things. The existence of dark energy might confirm

Einstein's cosmological constant, or it might signify the existence of another form of energy like quintessence. Whatever the outcome, the implications for the fate of the universe are big. But because the nature of dark energy is not yet known, what it potentially says for how the universe ends should be accepted only gingerly, as there is so much uncertainty involved with this brand new area of cosmology.

One thing we can do is to understand the implications of the accelerating universe on our view of the cosmos over time. The universe is expanding, and the space between galaxies is expanding, as time rolls on. But objects within the universe that are made of baryonic matter, like planets, stars, galaxies, trees, cats, and dogs, are not expanding over time. (Well, some of us may expand over time, but only if we eat too much.) So over long timescales in the cosmos, the distances between galaxies in groups and clusters, held together by gravity, will not change. The distances between objects inside a galaxy like the Milky Way will not change due to the accelerating universe. But because distant objects are being separated from each other more and more as time goes on, we will ultimately see less and less of the distant universe in the future.

The concept of what we can see in the cosmos is termed the horizon, named after what we can see on a planet like Earth. When I'm standing in my yard in Wisconsin, I can't see Paris. I have a limited horizon. And in the universe at large, we cannot see the most distant objects that exist either. And that problem will increase over time as objects are pulled increasingly far apart from each

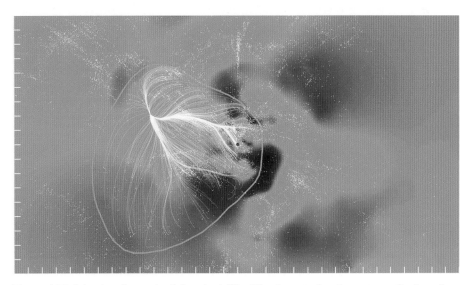

Figure 16.3 Scientists determined that the Milky Way is part of an immense gathering of galaxies they have named the Laniakea Supercluster, meaning "Immense Heaven" in Hawaiian. The unimaginably large formation holds the mass of 100 million billion suns inside a collection of some 100,000 galaxies.
SDvision interactive visualization software by Daniel Pomarède at CEA/Saclay, France

215

other. The accelerating universe will carry distant objects away from us more rapidly over time. Objects not bound by gravity will separate increasingly quickly in the future. Astronomers believe that dark energy is an intrinsic property of the universe at large, the fabric of space-time, and doesn't vary locally or regionally over the universe. If this is so, as space-time expands, there is more dark energy created along with it, and a runaway process ensues, with more space-time and more dark energy.

If this is so, then consider what gravity can and can't do. Gravity will become less important over time, as there is more space between objects in the cosmos. Gravity will not be able to overcome the vast gulfs between galaxy groups and clusters, and chains of galaxies will become more and more isolated as time goes on in a larger and darker, more isolated, emptier soup of the cosmos. Islands of normal matter – galaxies and the stuff they contain – will become far more isolated from each other, and the universe will become a darker, lonelier, more isolated place, with vast amounts of space between the bright matter that still exists. It is a troubling scenario for astronomy enthusiasts. Eventually, there will be far fewer galaxies for occupants of any given planet to observe, and the universe will no longer appear to be as populated with bright stuff as it really is, due to the incredibly large distances involved.

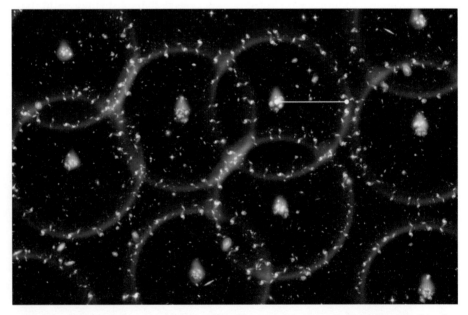

Figure 16.4 Galaxies have a tendency to cluster in spheres – seen greatly exaggerated in this artist's illustration – which scientists believe originated as waves in the early universe. This sphere's radius (white line) is a standard ruler that astronomers use to make precise measurements of the distances to galaxies, providing accurate predictions about the structure and evolution of the universe.
Zosia Rostomian/Lawrence Berkeley National Laboratory

And the apparent disappearance of the cosmos will begin to happen sooner than you might think. Over the next few billion years, we'll begin to lose sight of the most distant parts of the universe as their photons, sent toward us, are in effect pulled backwards by the ever-accelerating expansion of space. In a little more than a hundred billion years from now, the universe will look markedly different to anyone left in the Milky Way Galaxy. At that point in time, the universe will stretch to the voluminous size of some 1,000 times larger than it is now. And then, even nearby galaxies like those in the Virgo Cluster will not be visible to us, as their light will be stretched past our universal horizon. Only the remnants of the Local Group of galaxies, and a few other nearby groups, including our own Milkomeda, will remain for any sentient beings to stare at. To them, the local universe will be the entire universe.

The story becomes even darker and more bleak as time continues to roll on. In a universe 1 trillion years from now, all the stars in galaxies will be gone, save for the incredibly slowly burning red dwarf stars that live for trillions of years. They will feebly glow from the burned-out bellies of galaxies like dim embers in a cosmic fireplace. The accelerating expansion of the universe will continue to stretch, and the growing amount of space will make it impossible to see distant objects. So a trillion or more years from now, the observational universe we are familiar with will no longer exist. Any sentient beings on planets who may be around will be citizens of their own local universes, unable to see beyond that which is very close to them. Space will be mostly drained of radiation and baryonic matter, and they will be isolated and with an extremely short horizon. Will any residents have a record of the past that would tell them what they are seeing? Without a knowledge of the universe's history, they may have a very warped view of the cosmos (no pun intended). The residents of some planet in a ghost galaxy may in effect be living in a visual "black hole," unable to see anything beyond an extremely short horizon. Universal acceleration will by then put observational astronomy out of business, except for seeing our own remnant of Milkomeda!

In the days before dark energy, the fate of the universe was a relatively simple question of essentially two possibilities: either the universe was dense enough to halt the expansion and eventually pull everything back inward, creating a "big crunch," or the universe lacked the density to halt the expansion and everything would slowly move outward, leading to a lonely, dark, icy, dead cosmos. The second alternative indeed seems to be the very likely scenario. But dark energy has thrown a complicating factor into the realm of possibilities.

Because so little is known about dark energy, including what it is and whether it changes over time in any variant ways, the hypothetical possibilities for the universe's end have grown over the past decade. That is, in terms of the details. The basic scenario stays pretty much the same – a distant, outflowing, cold and dark universe. But the details can change significantly depending on the nature of dark energy. If it is believed to be Einstein's cosmological constant, the simplest explanation (one that would make cosmologists sleep reasonably well at night), meaning that dark energy is the same everywhere in space and time, then we could understand the universe's fate pretty well.

But dark energy as observed doesn't agree at all with what astronomers would expect the cosmological constant to be in standard models of particle physics. In fact, it is many orders of magnitude smaller than they would expect. This annoying discrepancy led to some cosmologists proposing quintessence, a dark energy force that is caused by a massive subatomic particle. In the quintessence model, dark energy can evolve from a very high value in the early universe to a much smaller value today, which would fit the observations more naturally.

Some cosmologists have taken the notion of quintessence to an even higher order by proposing an energy called phantom energy. This hypothetical form of dark energy would be a highly potent way to increase dark energy's effect on the cosmos beyond the simple concept of the cosmological constant. If it existed, in fact, phantom energy would cause the universe to expand to an infinite size in a finite time, producing an acceleration of the universal acceleration. The expansion would surpass the speed of light even more furiously than with normal dark energy, both of which are OK with Einstein because we're talking about the universe itself expanding, not particles within it. Would the difference in the acceleration from phantom energy make a difference? No one knows, as it's just an idea.

If the phantom energy scenario is true, then the universe is headed for a so-called Big Rip. What astronomers mean by this is that billions of years from now, the expansion of the cosmos and the growing strength of dark energy would begin to unbind objects now held together by gravity. That is, clusters and groups of galaxies would be ripped apart, shredded by the excessive force of whatever dark energy is. The Big Rip itself would occur some 22 billion years from now, when the fundamental forces in the universe, acted on by increasing dark energy, would come unglued.

The Big Rip hypothesis began its life in a paper published by American theoretical physicist Robert Caldwell of Dartmouth College and his colleagues in 2003. Taken with the physicists' most likely case, some 60 million years before the Rip, gravity would become too weak to hold individual galaxies together. Milkomeda, which will by then have been a merged galaxy for billions of years, will begin to shred apart. Some 3 months before the Big Rip, the physicists calculate, whatever is left of our solar system, along with countless other solar systems in countless galaxies, would be torn asunder. In the final minutes of the Rip, stars and planets would be gravitationally unbound, and very close to the final instant of the Rip itself, atoms would come apart and matter would exist no more.

The Big Rip scenario is highly unlikely given what we know about dark energy. It violates certain physical principles that are likely to be true. But we are in the realm of possibilities here, and not empirical observations. So there are certainly other possibilities that may happen too.

Most cosmologists would suggest that another scenario, the Big Freeze, is where the universe is heading. This is a little more mainstream approach, given what we know about dark energy. Most cosmologists believe the universe will expand forever and that a colder, darker, lonelier cosmos lies in the deep future. As billions and even trillions of years roll on, redshifts will stretch photons into undetectable wavelengths and eventually the supply of gas that could make new stars will be exhausted. Ultimately, stellar remnants will also be gone, leaving behind only black holes, which themselves will ultimately disappear due to Hawking radiation. An incredibly long way down the line, the universe will reach a point of inactivity called heat death.

The mileposts along the way through a Big Freeze scenario give us a fascinating glimpse of the probable future. We know the universe is now 13.8 billion years old, that our galaxy is some 9 billion years old, and the Sun and solar system are 4.6 billion years old (the thin disk; the globular clusters and halo are older). While these are incredible spans of time to humans, they are reasonably trivial in the total age history of the cosmos. The current Stelliferous Era, when normal stars and galaxies are operating as they should be, has a long time to go. As we have seen, some 4 billion years from now, the Milky Way and Andromeda galaxies will merge into a single galaxy we call Milkomeda. Some 100 billion years from now, the Local Group will merge into a giant, single galaxy, and 150 billion years from now all other galaxies will pass beyond our horizon. Communication between galaxies will then no longer be feasible.

The deep picture of the universe's future becomes even more bleak and lonely as time rolls on. Galaxies beyond the Local Supercluster, also known as the Virgo Supercluster, all the galaxies we see within some 110 million light-years of home, will appear redshifted to the degree that their light will have longer wavelengths than the observable universe, and thus they will become invisible to us.

Some 100 trillion years from now, the so-called Degenerate Era will begin when all star formation will have ceased. For a long time, the principal survivors in the otherwise burned out galaxies will be the longest lived stars in the cosmos, red dwarfs, which have masses around one-tenth of the Sun or less, and which live on for some 10 trillion years. When they become degenerate, they will be cooler white dwarfs. The universe will eventually be populated by white dwarfs and brown dwarfs, along with higher mass degenerate objects like neutron stars, as well as black holes. Most of the mass in this murky collection of objects will be tied up in white dwarfs, which will continue to become fainter over time.

The cosmos will by now be bathed in darkness, with only an occasional glimmer of light produced when a rare supernova erupts, caused by the merger of two white dwarfs. For a few weeks, a little diamond of light will flash on, a pinpoint in an otherwise universal sea of inky blackness. Planetary systems that may still exist around their degenerate stars will also decay over time as orbits fall inward or events occur that fling planetary bodies away from their stars. As

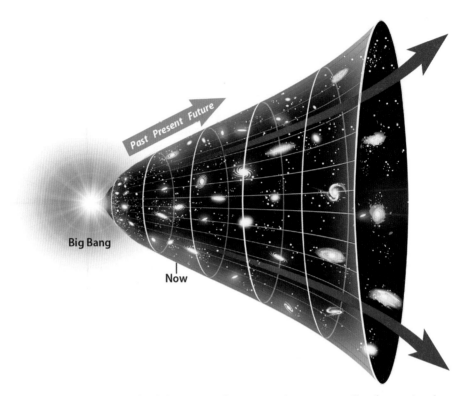

Figure 16.5 The latest models of the universe have it carrying on expanding forever in what astronomers call the Big Freeze, where everything in the cosmos gets increasingly far apart until galaxies outside our local supercluster disappear into the distance. In 100 trillion years, star formation will stop and the cosmos will be left with the slowly dying remains of dwarf stars.
Astronomy: Roen Kelly

we look at periods of even greater time, of 10^{19} to 10^{20} years from now, energy in objects still within galaxies will relax in an exchange of kinetics, causing lighter objects to gain kinetic energy and heavier objects to lose it. This means that many small objects will ultimately be ejected by galaxies, making what remains of the star systems smaller and denser. Ultimately, objects still within the burned-out galaxies will fall into their central supermassive black holes.

After some 10^{34} years, nucleons in the universe will begin to decay. This is uncertain, but the probability is that protons and neutrons may be unstable and have half lives of something like 10^{34} years. If this is so, then beginning around this time and continuing onward, much of the baryonic matter in the universe will be converted into gamma ray photons and leptons through the process of proton decay. Also assuming that nucleons decay, then by 10^{40} years, in effect, baryonic matter – protons and neutrons – will have decayed into photons and leptons.

After 10^{40} years, what remains of the universe will enter the Black Hole Era, wherein black holes will be the dominant, remaining citizens of the cosmos.

Though they will be the universe's dominant features, black holes will slowly evaporate over time through the process of Hawking radiation, which will leak matter away from them. Because of Hawking radiation, a 1 solar mass black hole will evaporate over 2×10^{66} years, and a supermassive black hole of 100 billion solar masses will decay over 2×10^{99} years. As the black holes evaporate, they leak matter mostly in the form of photons and probably hypothetical gravitons. As they lose mass, black holes increase in temperature, eventually warming to the point where the hole will produce a small amount of light, another short glimpse of illumination in the otherwise black cosmos.

Beyond the Black Hole Era, to a day some 10^{100} years from now, the universe will enter the Dark Era. Although the universe will by then be mostly empty, particles will be moving around, rarely encountering each other. The universe will be gravitationally controlled by dark matter, with electrons and positrons also playing significant roles by their presence. The universe will by now have reached an extremely low state of energy, with interactions between particles coming few and far between. It will not only be a very dark universe, but one that is essentially inactive.

No one as yet knows what lies beyond this dark, sleepy, distant future of the cosmos. The universe may stay this way nearly forever. Some theoretical physicists have suggested it may undergo another period of cosmic inflation, or that a Big Rip scenario may take place. Could it be that the macro-scale physics we know so well would break down completely and leave the universe in a state controlled by quantum physics? Or that the current vacuum state of the cosmos is a false vacuum, which could decay into a lower energy state? No one yet can say with any degree of certainty what will happen to the universe such a long time from now.

It does appear that ultimately the universe will undergo so-called heat death, becoming not only dark and dead, but also eternally cold. In this state, no thermodynamically free energy would be available and so no more processes could occur that would consume energy. It would be the ultimate, authoritative ending for any activity in the universe. What an anticlimactic way to end it all!

We have already teased onto the subject of another possible fate of the cosmos, the Big Crunch. This would see the expansion of space reversing and everything falling back onto itself, for another epochal Big Bang. Whereas this idea has long origins and supporters years ago, it has lost likelihood in the minds of nearly all cosmologists today. Although a Big Crunch scenario seems highly unlikely, there is a wild card in the mixture. The nature of dark energy, largely unknown as yet, could offer a possibility of the Big Crunch being an outcome in the universe if the dark energy someday becomes gravitationally attractive and if there's enough of it.

The fate of the universe raises an intriguing, related, scientific-philosophical question about existence. Is the universe we know of and exist in the only universe that could exist? Could other universes exist beyond our own? This

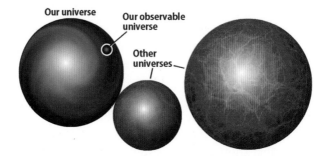

Figure 16.6 A significant number of cosmologists believe that – in the same way our ancestors have been astonished to find other planets, stars and galaxies – our universe might not be the only one. Theoretical physicists are drawn to the idea because it might explain gaps in current understanding of the cosmos, but proof could be impossible.
Astronomy: Roen Kelly

question has been a purely philosophical one for a very long time, and physicists have invoked various cosmological ideas to suggest that we may live in a "multiverse" – that our universe we can see and we exist in is one of several or perhaps infinitely many universes that exist. The potential logic behind this thinking comes entirely from theory and from mathematical possibilities – of course, there is as yet no observational evidence of other universes beyond the one we know, and by the nature of empiricism, there may never be evidence for other universes, even if they exist. So the matter, for the present at least, is a hypothetical and philosophical one rather than a pure scientific question.

Nevertheless, some important cosmologists, including Stephen Hawking, Steven Weinberg, Max Tegmark, Alan Guth, Andrei Linde, and others, support the possibility or plausibility of multiple universes. The philosophy of different universes, if you will, could mean somewhat different things, and so the Swedish–American cosmologist Tegmark (1967–) of the Massachusetts Institute of Technology, has devised a classification scheme to enable understanding what physicists mean by the phrase "other universes."

Tegmark's first level of another universe represents a cosmos beyond our cosmological horizon – that is, related to cosmic inflation, parts of the universe would have regions called Hubble volumes surrounding any given spot in which objects are receding from observers faster than the speed of light, due to universal expansion. Most such Hubble volumes in the cosmos would have different conditions than ours (although they would be governed by the same physical laws), but because of the possibly infinite size of the cosmos, Tegmark argues, an infinite number of Hubble volumes could be identical to ours.

Second, inflation theory suggests the universe (and other universes) stretch over time, and bubbles form between portions of the multiverse, which would form in effect new universes. Different so-called bubbles forming in the expanding multiverse may have completely different physical laws or constants than the universe we are familiar with.

Third, quantum mechanics enters the fray and suggests that a so-called "many-worlds interpretation" could be real. This suggests that many possible observations of events in a multiverse, each having a different probability, correspond to a different universe. For many years, the apparent randomness in nature bothered scientists. Albert Einstein famously rejected this uncertainty and summed up his feelings with the phrase, "God does not play dice." More recently, however, Stephen Hawking has discussed quantum mechanics and the uncertainty of nature. In discussing particles falling into a black hole, for example, Hawking describes how the properties of such a particle cannot be measured. "Not only does God definitely play dice," says Hawking, "but He sometimes confuses us by throwing them where they can't be seen." The uncertainty wrought by quantum mechanics means that in a many-worlds interpretation of the universe, other "parallel" universes could exist in other dimensions of the universe we inhabit, or in bubble universes created during cosmic expansion.

Tegmark himself suggests the possibility of multiple universes existing with different mathematical properties.

The implicit possibility of other universes existing because they mathematically could exist is strongly suggestive for some cosmologists. Other astrophysicists don't believe in multiverses because their existence simply cannot be observationally tested, and therefore, by definition, they simply lie outside the realm of testable science. At least for the time being. At an absolute minimum, however, they give us amazing material to discuss at cocktail parties.

Chapter 17
The meaning of life in the universe

Nothing drives astronomy like that oldest of all philosophical questions: Are we alone in the universe? Spacecraft missions concentrate on Mars because of the Red Planet's relative similarity to Earth and the existence of water there, leading to the possibility of microbial life. The rapidly growing cottage industry of finding and studying extrasolar planets looks forward to detecting Earth analogs that may also reveal atmospheric signatures of living beings. The discovery of life elsewhere in the cosmos would certainly mark one of the most incredible moments in human history, a milestone at which we would understand we are not unique in the universe.

Of course, we know of only one example of life in the universe, right here on Earth. In the minds of some, that means the odds of life being an extremely rare thing in the cosmos are high. At least intelligent life; civilizations that could communicate. They point back to the idea that Italian physicist Enrico Fermi (1901–1954) raised in 1950: "If the universe contains life, then where is it? Why hasn't life showed up on our doorstep?" The so-called Fermi Paradox still stands as a fair question. But the odds of life in the universe are staggeringly large, overwhelmingly so, in the minds of the majority of astronomers and cosmologists.

Recall that the universe contains at least 100 billion galaxies, and probably considerably more because inflation theory means we are not seeing the whole universe that exists. And let's consider the number of stars in a galaxy like the Milky Way, about 400 billion. Let's set inflation aside. From what we see of star systems near the Sun, planetary systems appear to be common, and we are seeing the first glimpses of planets within the habitable zones of their suns – the areas in which water would be a liquid. From what we can see, water is essential for life.

Astronomers currently believe that something like 70 or 80 percent of stars have planets. With 100 billion galaxies in the universe and to play it conservatively, let's assume roughly 100 billion stars on average for each galaxy. That works out to 10,000 billion billion stars in the universe, and roughly 10^{22} planetary systems, or 8,000 billion billion.

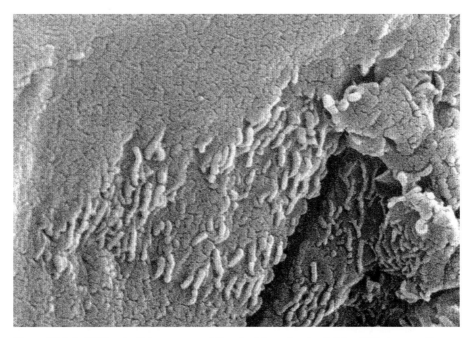

Figure 17.1 In 1996, scientists announced they had found potential fossil life in a martian meteorite, Allan Hills 84001 – but the "life" turned out to be highly controversial. NASA/JSC/Stanford University

Do we really believe that we are the only planet on which life exists? Or the only planet on which a civilization exists with so-called intelligent life? Even simple logic and probability suggests that the odds seem mind-numbingly against that.

But before examining the meaning of life in the universe, we need to begin by understanding just what life is. Defining life has proved to be an intriguing and difficult task over the ages. Early materialists like the Greek philosopher Empedocles (ca. 490 BC–ca. 430 BC) suggested that everything in the universe is made of a combination of four elements: earth, water, air, and fire. Changes in substance were to be explained by rearrangements of these four simple elements. By contrast, Greek philosopher Aristotle (384 BC–322 BC) believed in hylomorphism, life consisting of a combination of matter and form, and that living beings got their form from a soul. More recently, German chemist and physician Georg Ernst Stahl (1659–1734) suggested a hypothesis of vitalism, in which organic and inorganic matter were fundamentally different, and that organic material can only be derived from living beings. But this ended in 1828 when German chemist Friedrich Wöhler (1800–1882) synthesized urea from inorganic materials, thus inaugurating the field of organic chemistry.

Modern definitions of life start with a simple idea: When you see something that's alive, you know it. But the underpinning of the definition of life is more complex, and consists of several attributes. First, living things display order, for example in the arrangements of atoms in their molecules. Lots of inorganic molecules also have orderly arrangements of atoms (think water, salt, etc.), but living things must have order, while nonliving things don't necessarily need to be ordered. Second, living things reproduce or are products of reproduction. Reproduction is a necessity of life, as without it, life would not go on. Third, life exhibits growth and development. Living things grow and develop over time in part governed by their heredity, which is in turn controlled by inherited DNA. (Although this might not be true for life elsewhere.)

Fourth, living organisms utilize energy from their environment, according to the second law of thermodynamics. Without energy from our environment, we would quickly die. Living beings take in energy and transform it in order to operate their bodies and counter entropy, which pushes everything in the universe from order to disorder. Life is, in large part, a fight against increasing entropy that we eventually will lose. Fifth, living things respond to their environment. Creatures may move around to find preferable temperatures or sources of food. Many aspects of the human body – sweating, blood circulation, hormonal changes, and so on – are responses to the environment around us. Rocks, by comparison, do not actively change in response to such environmental effects.

Sixth, living beings show evolutionary adaption based on Darwinian ideals. Over time, life on planet Earth has altered dramatically due to interactions between individual organisms and the environment. Traits that allow individuals to survive over time mean the survivors pass on their genes to subsequent generations. This is how living beings achieve "immortality" – by passing themselves on to subsequent generations. Insects that happen to appear camouflaged because they resemble tree bark are less likely to be eaten by predators and survive to reproduce again, shifting the surviving traits in later generations. Plants that happen to produce toxic substances are eaten less frequently than other plants, and so pass on their genes through time. Gradually, over the billions of years of life on Earth, Darwinian evolution shapes and changes the character of living beings so that certain species live on more successfully than others.

The history of life on Earth has left us a rich heritage. The earliest known life comes from the Strelley Pool area of Western Australia, as we have seen, and dates to about 3.4 billion years ago. These are primitive cyanobacteria, tiny spherical, oval, and tubular structures. Other researchers have found life claimed to be as old as 3.9 billion years, but these results are as yet unconfirmed.

Clearly, life on Earth took hold relatively quickly following the so-called Late Heavy Bombardment some 4.1 to 3.8 billion years ago, the period in which many small bodies in the solar system slammed into Earth and other inner planets and moons. The development of life on Earth, driven by Darwinian evolution, leaves a fascinating timeline behind and makes us wonder if very similar evolutionary paths may have occurred on countless other planets in the universe.

The origin of life on Earth has long been sought, and scientists are just now getting a handle on how it probably happened. Clearly the constituents of important biotic molecules exist in interstellar space, and even complex organics like amino acids have been recovered in meteorites. Some 4.2 billion years ago, Earth had a stable hydrosphere, and soon thereafter prebiotic chemistry. Around 4 billion years ago, chemists believe, a pre-RNA world existed leaning toward the creation of nucleotides that would ultimately make up RNA and DNA, the two most important biological molecules. Membrane-forming molecules were also needed, to form cellular like structures that would enable the chemistry of life. Experiments suggest these were relatively easy to produce in the prebiotic soup of Earth's oceans, and it may be that the origin of life occurred deep within the oceans, alongside so-called black smokers, hydrothermal vents rich in the right chemistry and enough heat to get the process of making complex molecules going.

Once self-replicating molecules got their start, life took off, but it remained simple for a very long time. Life evolved in step with the major geological eras of the planet, as well. At first, life was extremely simple, so-called prokaryotic,

Figure 17.2 The celebrated meteorite Allan Hills 84001, recovered in Antarctica in 1984 and, 12 years later, the center of a major scientific controversy when a team of scientists claimed to have found fossils left by microbial life in the rock. (Virtually all scientists have not accepted this notion, and believe that the micro "fossils" were imprints of large chemical molecules.) The meteorite resides in the collection of the Smithsonian Institution's National Museum of Natural History, where the author photographed it in its custom-made container in the spring of 2014.
David Eicher

consisting of a single-celled bacterium with a simple cell wall. For 1.5 billion years, organisms on Earth consisted of these extremely simple types, with organisms like cyanobacteria producing photosynthesis, garnering energy from the Sun. Some 2 billion years ago, more complex cells, eukaryotes, evolved, containing not only cellular walls and a nucleus but also more complex organelles. Multicellular life appeared on the scene about a billion years ago.

Some 570 million years ago, animals emerged, and soon thereafter the so-called Cambrian explosion took place. In warm, shallow seas, and with plentiful atmospheric oxygen, Earth now made a hospitable home for complex living beings. The earliest vertebrates appeared, along with a wide variety of creatures, including fungi, sponges, and corals. Around 520 million years ago, the climate warmed in the Ordovician period, wherein extensive, shallow seas hosted marine invertebrates, and jawless fish became widespread. The Silurian period, some 450 million years ago, witnessed the first jawed fish and terrestrial arthropods.

Some 420 million years ago, the Devonian period featured a cool climate and the development of freshwater basins. During this time, fish diversified and the first amphibians appeared. The warm, humid Carboniferous period, some 375 million years ago, created extensive coal-producing swamps, the first reptiles, and plentiful arthropods and amphibians. Some 285 million years ago, the intriguing Permian period featured a cool climate early that warmed throughout, and common reptiles, along with amphibian extinctions.

As time rolled on, life continued to become more complex, driven by Darwinian evolution. The Triassic period, 240 million years ago, featured a warm climate and extensive deserts, with dinosaurs replacing the reptiles. The first true mammals also appeared on the scene. During the Jurassic period, some 195 million years ago, the climate was warm and stable, and the first birds appeared, along with a great diversity among reptiles. The Cretaceous period, beginning some 135 million years ago, saw continental seas and swamps spread across the planet, and the extinctions of ancient birds and reptiles. During the Paleogene period, beginning some 66 million years ago, enormous diversity in mammals pressed forward, including the rise of the first hominids, our earliest ancestors, some 6.5 million years ago.

Two million years ago the first members of the genus *Homo* appeared on the scene; more recent ancestors appeared in the forms of *Homo heidelbergensis* (600,000 years ago), Neanderthals (300,000 years ago), and, finally, anatomically modern humans, some 200,000 years ago in Africa. Some 50,000 years ago modern human beings, *Homo sapiens*, began inhabiting other continents, replacing Neanderthals in Europe and other hominids in Asia. About 30,000 years ago the last remaining Neanderthals died off, and for the past 10,000 years the Holocene epoch on Earth commenced with modern humans living in a post-Ice Age world, complete with an evolution from hunting and gathering to settled agriculture and the first cities on planet Earth. And here we are.

The role of extinctions in the evolution of life on Earth should also be well remembered, as it has played an important part in the story of life. As we have seen, many large bodies have impacted Earth since the planet accreted. The largest such impact is believed to be the Theia–Earth collision early on that led to the formation of the Moon. Other, smaller impacts leave tourist sites, as with the Barringer Meteor Crater near Flagstaff, Arizona, a 1,200-m (3,900-foot) scar dating back 50,000 years. We see an abundant and well-preserved record of impacts on the Moon, but Earth, with its many resurfacing processes, hides evidence of past impacts. Nonetheless, scientists have uncovered evidence of them.

The most famous of all impacts is, of course, the so-called K-Pg Impact (short for the Cretaceous-Paleogene), which occurred 66 million years ago, at the end of the Cretaceous period and the commencement of the Paleogene. (This was formerly called the K-T Impact, after the Cretaceous and the now-obsolete term Tertiary.) A 10-kilometer asteroid slammed into the Yucatán Peninsula, and the resulting firestorm and nuclear winter scenario killed off much of life on Earth, including the dinosaurs. Fortunately, for us, small, squirrel-like mammals survived, and we owe our existence to that fact. Some 75 percent of all species vanished.

Other major extinctions have also occurred, however, and the cause for most is not clear. Going back in time, the Triassic–Jurassic extinction event some 200 million years ago witnessed the extinction of 70 to 75 percent of species. The Permian–Jurassic extinction event, known as the "Great Dying," happened some 251 million years ago and saw the end of 90 to 96 percent of all species. The Late Devonian extinction event, 360 to 375 million years ago, saw the end for 70 percent of all species. The Ordovician–Silurian extinction event, 440 to 450 million years ago, was the end for 60 to 70 percent of species in existence at the time.

Scientists have a pretty good understanding of the history of life on Earth. But what about the existence of life out in the universe, among those other 10,000 billion billion star systems? The first and easiest place to look for life in the universe is our own solar system, and there are plenty of worlds right here at home, with planets orbiting the Sun, moons encircling planets, and many smaller bodies – dwarf planets, asteroids, and comets. Our understanding of the way self-replicating molecules work suggests that water – or another, similar solvent – is a major, necessary factor in living organisms. This is certainly the case on Earth and may well be the case throughout the universe, or in other cases another similar solvent would play the role water does with us.

Several intriguing places in the solar system offer watery environments, or at least places where once abundant water existed. This is, of course, the case with Mars, as we have seen. Mars has plenty of water ice still, and likely subsurface aquifers. So could life exist on Mars today, or did it in the past? Of course, by life in the solar system, we're talking about microbial life.

Several other places in the solar system offer a compelling possibility for the existence of microbial life. Planetary scientists believe Jupiter's moon Europa has an extensive liquid water ocean beneath its icy crust. The sixth-largest moon in the solar system, Europa is rich in silicates and probably has an iron core, a tenuous atmosphere, and an icy surface striated by cracks and faults. The extensive amounts of water ice on Europa and tidal flexing help to create the subsurface ocean, the existence of which was bolstered in 2014 with the detection of plate tectonics on Europa's thick ice crust, the first detection of this activity on a planetary body other than Earth.

Further, in 2013, NASA scientists detected phyllosilicate clay minerals on Europa's surface – and these minerals on Earth are often associated with organic molecules. The moon also displays evidence of water vapor plumes similar to those of the saturnian moon Enceladus. Most of what astronomers know about

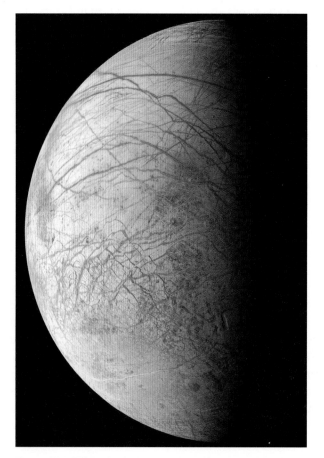

Figure 17.3 When the Voyager spacecraft returned the first images of Jupiter's moon Europa, Carl Sagan suggested cracks in the ice could be promising for finding oceans and maybe even life.
NASA/JPL/Ted Stryk

Europa is derived from observations by the Galileo spacecraft, which was launched in 1989 and entered a jovian orbit in 1995. The European Space Agency is planning a mission to investigate Europa and jovian moons Ganymede and Callisto for potential signs of life. Ganymede and Callisto also show some promise of harboring subsurface oceans, but the energy dynamics of their systems make it somewhat less likely that they would contain life. The spacecraft, named JUpiter ICy moons Explorer (JUICE), is slated for a 2022 launch. NASA also hopes to launch a similar mission several years later.

Another possible abode for microbial life in the solar system is the saturnian moon Titan. Saturn's largest moon – and the second largest moon in the solar system – Titan is the only moon to have an extensive atmosphere. Titan's volume is larger than that of the planet Mercury, although its density is very much lower. Titan is a world consisting of rocky materials and abundant ice, and its atmosphere is mostly nitrogen. The revolutionary Cassini–Huygens mission studied Titan extensively in 2004, discovering many liquid hydrocarbon lakes on the moon. The spacecraft's Huygens component landed on Titan, providing valuable science and exciting images from the moon's surface. Lots of liquid methane and ethane exist on Titan, as well as extensive clouds. The surface contains abundant rivers, lakes, seas, and dunes, and Titan experiences a weather cycle like that of Earth, except that methane rains out of the sky. (The temperatures on Titan are far too cold for liquid water, of course.)

In 2014, NASA scientists announced that Titan's extensive nitrogen atmosphere likely originated from the Oort Cloud, the reservoir of 2 trillion comets on the solar system's outer edge. They also suggested that a subsurface ocean in Titan could be extremely salt-rich, akin to the Dead Sea on Earth. They also reported that the methane rainfall on Titan could interact with icy materials underneath the moon's surface, producing ethane and propane that flows into lakes and seas on Titan's surface.

Two other moons have revealed evidence for subsurface liquids, and they are Saturn's moon Enceladus and Neptune's moon Triton. Enceladus is Saturn's sixth-largest moon and is a water-rich world, an icy globe only about one-tenth the size of Titan. Years ago, the Voyager 2 spacecraft revealed that Enceladus has a wide range of surface features, from tectonically deformed areas to heavily cratered regions. Our knowledge of Enceladus grew rapidly in 2005, when the Cassini spacecraft explored it extensively. Cassini discovered water-rich plumes emanating from the moon's surface, cryovolcanoes shooting geyser-like fountains of water vapor, ice particles, salts, and other volatiles into space. More than 100 geysers on Enceladus have been identified. Not only is Enceladus geologically active in unusual ways, but in 2014 NASA scientists announced their discovery of a large south polar subsurface ocean using Cassini data.

Even farther out, at much colder temperatures, astronomers have hopes of someday exploring oceans underneath another moon's icy crust. In the Neptune system, some 30 times farther away from the Sun than Earth, is Triton, the

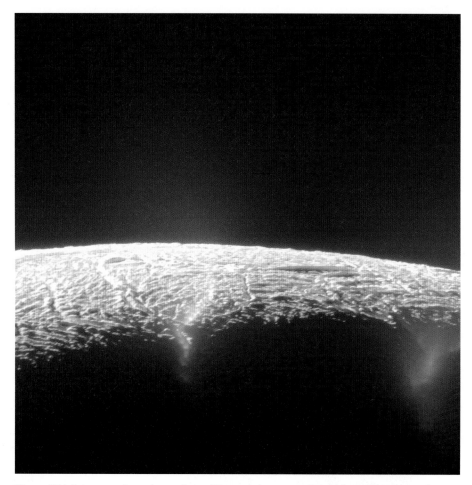

Figure 17.4 Jets erupt from the surface of Saturn's icy moon Enceladus in this image from NASA's Cassini spacecraft. Astrobiologists think worlds like this might potentially harbor simple life forms.
NASA/JPL-Caltech/Space Science Institute

planet's largest moon. Triton is the seventh largest moon in the solar system. This strange world is a geologist's dream: its surface consists mostly of frozen nitrogen, with a significant water-ice crust, and oddball, fractured, "cantaloupe terrain," showing signs of stress fractures and buckling. Geologically active, Triton's surface is young, and shows evidence of cryovolcanism, as with Enceladus. Geysers on the moon erupt with showers of liquid nitrogen, and Triton also has a tenuous nitrogen atmosphere. Planetary scientists have evidence of subsurface water and ammonia on the strange moon.

Whether or not life exits in our solar system – aside from right here – we ought to know something about how to search for it further afield, too. The nature of where and how to look for life in the larger universe is a very tricky

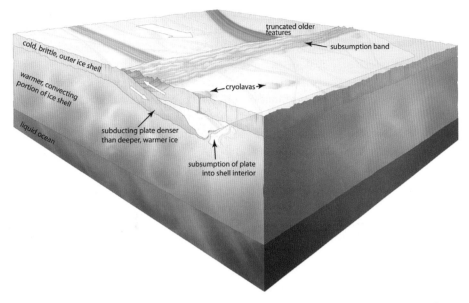

Figure 17.5 In 2014, astronomers found evidence that a form of plate tectonics was pushing Europa's ice into massive collisions. This subduction could feed nutrients from the surface to any potential organisms living in the ocean below.
Astronomy: Roen Kelly

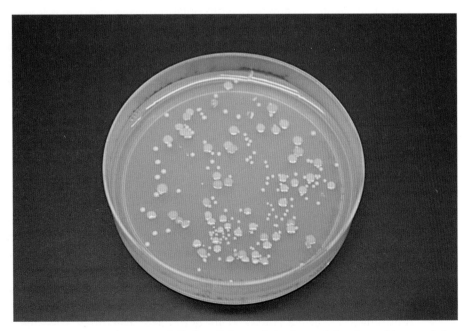

Figure 17.6 A team of scientists drilled a mile beneath the Antarctic ice sheet with excruciating care and were able to culture bacterial colonies from Lake Whillans water column samples. The extremophiles show that life is possible even where the Sun's rays can't reach.
Louisiana State University

one. Where to search? As we have seen, hundreds of stars near us in the Milky Way have planetary systems, which is no surprise given what astronomers believe about the formation of stars and the attendant debris surrounding them. But which star systems would we try to explore?

Planetary scientists have long admired the concept of the habitable zone. We know about the importance of liquid water for life. And we know from our own planet about harboring a comfortable range of temperatures not only for the existence of liquid water but also for the functioning of all sorts of biological processes. Earth orbits our Sun between distances of about 147.1 million kilometers in wintertime (Northern Hemisphere) and 152.1 million kilometers during midsummer. The variation in distance has essentially no effect on weather, but it reminds us that there are distances too large for comfortable life on Earth – and too small, too.

The so-called habitable zone, or goldilocks zone, is defined by the region for a given star in which a planet will not be too cold nor too hot for the existence of liquid water. The concept is used by planetary scientists to define where earthlike worlds could exist. But, of course, being in this zone is not necessarily strongly suggestive of supporting life – the Moon is in our star's habitable zone, and it's as dead as a doornail. Moreover, habitable zones evolve over time as their host stars evolve – with sunlike stars, the habitable zones would slowly push outward over time as the stars brighten. Additionally, planets or moons outside a conventional habitable zone could contain life, due to internal oceans, and so on, as might be the case with outer moons in our own solar system.

But defining a habitable zone is a tricky business. Even the zone of our own star is not exactly clear. The most optimistic estimates for our own habitable zone place it between 0.84 to 1.7 astronomical units, from midway between the orbits of Venus and Earth to outside the orbit of Mars. A conservative estimate suggests it is confined within 0.95 and 1.4 AU, just inside Earth's orbit and most of the way out to Mars.

We have already witnessed the extraordinary growth of extrasolar planet discoveries over the past 20 years. More than 2,000 planets orbiting stars other than the Sun are now known, and these represent searches generally very close by in the galaxy and over small areas of our sky, too.

Detecting life on planets from afar would be a difficult and tricky business. But much less so than building spacecraft and attempting to visit them over the breathtakingly large distances between even the nearest star systems. Astronomers have already detected Earth-sized planets around other stars. Soon, they will have ideas about the possible habitability of some of these worlds.

If the new generation of telescopes can provide a direct image of a distant earthlike world, even if it's only a few pixels across, we could learn something from any possible variations in its brightness over time. This might reveal something about the reflectivity of its oceans relative to landmasses, for example. A simple spectrum of the planet could reveal the planet's rough surface temperature, or perhaps even a spectral analysis that would allow guessing whether or not the world could support life. Clues about the planet's ratio of rocks to ices might appear. Infrared spectrometry might allow a rough estimation of the atmosphere of a very distant planet – how much carbon dioxide, ozone, water vapor, or methane is present. Measurements like this would be suggestive of the existence of life, if only based on their abundances.

This leads to another question: How rare are other Earths in the Milky Way, or in the universe at large? Some Earth-sized planets might be earthlike too, given what we know about the accretion of planets around stars – possibly both in composition and in terms of relative distances from their host suns. But some scientists strongly contest this, describing the so-called Rare Earth Hypothesis, that states the "incredibly unlikely chain of events" that would all have to happen to lead to complex life on a planet. They also suggest that life on an earthlike world could only exist in a relatively small region of the galaxy, leading to the idea of a galactic habitable zone. According to this idea, outer regions of a galaxy would be unlikely to host life because of too-small amounts of elements heavier than hydrogen and helium. Inner regions of a galaxy could be ruled out due to high rates of supernovae that could kill nearby life. According to the Rare Earth Hypothesis folks, perhaps only 10 percent of the volume of a galaxy's disk could host life, at least on an ongoing basis.

But there are strong counterarguments to these ideas, too. In terms of heavy elements, scientists advocating this view argue that Earth's mass is about three one-hundred-thousandths of the Sun's, and so actually a small amount of heavier elements is all that's needed for earthlike planets. Scientists also point out that the radiation danger from supernovae, as far as its effects on life go, is speculative. The radiation could even help life by creating mutations that would help the journey along Darwinian evolution!

The effect of impacts from small solar system bodies on the long-term viability of life is another issue. In our solar system, the Late Heavy Bombardment probably made life untenable on Earth from about 4.1 to 3.8 billion years ago. Perhaps periods of prolonged bombardment would characterize other solar systems. In our solar system, Jupiter has protected us gravitationally, sweeping up and engulfing or casting away many small bodies. Solar systems that do not have a Jupiter would not afford this protection. Moreover, the stability of a planet's climate could be an issue to long-term survival. Earth's climate stability derives in part from plate tectonics, which regulates the carbon dioxide cycle. The existence of the Moon also plays a role because it means that Earth's axial tilt varies only a small amount.

So we don't really know how common other Earths might be in the Milky Way or in the universe at large. Each argument from one side faces a counterargument from the other. But astronomers are beginning to learn lots more about other planets and that will help open our eyes to the truth.

As we have seen, the enormity of the cosmic distance scale makes visiting other life in the universe only a remote possibility. But detecting life in the universe, either through spectra from other worlds or the electromagnetic radiation from civilizations, is another matter. Consider Earth: for more than 50 years, we have produced high-frequency radio, radar, and television transmissions. Roughly 2,500 stars lie within 50 light-years of Earth. Conceptually, any civilizations within this radius could have detected our presence through *I Love Lucy*, *The Twilight Zone*, the broadcast of the Senate Watergate Hearings, or a million other things.

Of course, no one as yet knows of life elsewhere in the universe, or even the chances of how widespread it may be – simple or complex – if it exists. Arguments wash back and forth between highly skilled scientific thinkers, some on the liberal side and others far more conservative. The first scientific attempt to quantify how many intelligent civilizations might exist in the universe took place in 1961, when American astronomer Frank Drake (1930–) presented an equation at a conference held in Green Bank, West Virginia discussing life in the universe.

The so-called Drake Equation specifies that the number of civilizations now transmitting an electromagnetic signal is equal to $N_{HP} \times f_{LIFE} \times f_{CIV} \times f_{NOW}$. (This four-term modification from the original seven-term equation, like the original, focuses on just the number of civilizations in our own galaxy, which can be multiplied by the number of galaxies to get an estimate for the whole universe. Moreover, signals from other galaxies would be weakened, making them harder to detect.)

The terms are straightforward: N_{HP} is the number of habitable planets in the galaxy, a number we're starting to get a feel for. f_{LIFE} is the fraction of habitable planets that actually contain life. f_{CIV} is the fraction of life-bearing planets on which a civilization capable of interstellar communication has arisen, at some point in its history, if not at present. f_{NOW} is the fraction of civilization-bearing planets that have an active civilization at the present time.

What good is the Drake Equation? It affords astronomers the opportunity of estimating the number of civilizations with the capability of communication in the Milky Way Galaxy and by simple multiplication the number in the entire universe. For example, suppose that 2,000 habitable planets exist in the Milky Way, and 1 out of 20 has life, 1 in 5 with life that at some time had an intelligent civilization, and 1 in 10 civilizations that has ever existed exists now. That means there are currently two civilizations in the Milky Way, and we're one of them. (Many astronomers would argue that these sample numbers, particularly the number of habitable planets, are a little on the conservative side ... ! Drake estimated the value of N at 10,000, and Carl Sagan figured a million or more.)

Astronomers are experimenting with radio and optical telescopes to search for other civilizations in the cosmos. These searches, collectively called SETI (for the Search for Extraterrestrial Intelligence), form an incredible aspect of modern astronomy, and much of the work is centered with the California-based organization called the SETI Institute, established in 1984 following a number of early SETI search experiments. The origins of modern SETI actually occurred at

Cornell University in 1959, when Italian particle physicist Giuseppe Cocconi (1914–2008) and American physicist Philip Morrison (1915–2005) wondered about sending radio signals over interstellar distances.

The current techniques of radio searching for civilizations owe much to Cocconi and Morrison. In 1960, a year before devising his famous equation, Frank Drake conducted Project Ozma using the 26-m radio dish at Green Bank, searching for telltale radio signals from two nearby stars, Epsilon Eridani and Tau Ceti. He did not detect alien signals, but this effort led to many others in the years that followed. In 1974, astronomers beamed a message to the Hercules Cluster (M13), the globular cluster that contains hundreds of thousands of stars.

Modern SETI can search for three types of signals – signals used for local communication, signals used for communication between a civilization's home planet and another colony or spacecraft, and intentional signal beacons sent into space so that others would discover them. Just how would detecting a signal work? In his famous book and the subsequent movie *Contact*, Carl Sagan suggested that other civilizations might play back one of our own transmissions to us. This would only work for very close neighbors! The more likely strategies for detection would be to look for repeating signals.

Messages between civilizations could also be encoded in ways similar to those used for Internet transmissions. But the almost countless ways of encoding messages might mean that genuine alien messages would be extremely difficult or impossible to recognize. Broadcasting civilizations really wanting to be found, however, would probably make their transmissions simple and easily recognizable.

Radio searches for civilizations are now conducted either by targeted searches using dishes focused on particular star systems, or in sweeping techniques called sky surveys. Current radio SETI projects include three surveys, SERENDIP, Southern SERENDIP, and SETI Italia. Additionally, the Allen Telescope Array at Hat Creek Radio Observatory in northern California has been in operation for close to a decade. The SETI Institute's experiments use an array and millions of channels. Their sensitivity is increasing over time, too, making the detection of alien civilizations a real possibility over large distances.

Most SETI searches have used radio telescopes, but a newer and developing area of research is so-called optical SETI, which aims to detect laser emissions that could encode information in interstellar transmissions. Visible light has the disadvantage of being absorbed by tiny dust grains throughout intergalactic space. Nevertheless, researchers believe optical SETI signals could operate across thousands of light-years. And millions of stars lie within 1,000 light-years of Earth.

The multiple efforts now underway to search for extraterrestrial civilizations invariably raise the favorite question of many TV shows: has alien life visited Earth in the form of UFOs? After all, half the American public believes alien beings have visited Earth. The claims of UFO proponents, when actually subjected to the principles of scientific analysis, are not very good. Moreover, anyone who spends a few hours investigating the distance scale of the universe should realize, with a fair bit of sense, that the odds of traveling around the galaxy, given the time and energy realistically required, are not promising.

Reports of UFOs commenced in earnest just after World War II with a well-publicized incident near Mount Rainier and the claims of a businessman who said he spotted nine shiny objects moving at high velocities. A reporter's misunderstanding led to the term "flying saucers." But the tens of thousands of UFO reports made since, many of which have been thoroughly and scientifically investigated, have yielded nothing about the existence of life elsewhere in the universe. They have revealed a great deal about the human beings here on Earth who have made the reports. People see things in the sky they often don't understand. That's the conclusion.

The fact that observers can't identify an object, or that it seems mysterious to them, should not be surprising – particularly given the nature of some of the reports. The "unidentified" part of the term UFO does not mean turning over all we know about science, about energy and the cosmic distance scale, about Occam's Razor, and leaping right into alien visitation. Far more solid evidence would have to be collected and analyzed by actual scientists to get over that hurdle. (Occam's Razor is the 700-year-old principle in logic that, among competing hypotheses, the one made with the fewest assumptions should be used and almost always turns out to be correct. It was created by the English Franciscan friar William of Ockham [ca. 1287–1347].)

Face it, folks, it's a very, very big universe. But that gives us a little perspective, if nothing else. Earth is a pretty special place, at least for our species, and we should take good care of it, our fellow human beings, and everything else that lives along with us. It is the only home we have.

Glossary

Accelerating universe   The observation that the universe is expanding at an increasing rate.

Accretion   The formation of astronomical bodies as material is drawn together by gravity.

Accretion disk   A flattened structure in orbital motion about a massive central body such as a black hole.

Active galactic nucleus   A compact region of the center of a galaxy with a very high luminosity and mass that is a highly energetic source of radiation; powered by a black hole centrally located with the galaxy's nucleus.

Adaptation   The ability of a biological organism to adapt to changing environmental circumstances, and to adapt via genetic changes.

AGN   See Active galactic nucleus.

Albedo   The degree of reflectivity of a celestial body.

ALH 84001   See Allan Hills 84001.

Allan Hills 84001   An Antarctic meteorite found in 1984 that was the subject of a celebrated 1996 study claiming the discovery of microfossils in this stone from Mars. Subsequent research showed the tiny structures to be non-biological.

Alpha-Proxima Centauri   The closest star system to the Sun, consisting of a close double star, Alpha Centauri (4.4 light-years away), and a small companion star also in orbit, Proxima Centauri (4.2 light-years away).

Altimetry   Measurement of altitude or distances relating to a spacecraft.

Amino acids   Biologically important compounds made from amine and carboxylic acid groups, and constituting the building blocks of proteins.

Andromeda Galaxy   The closest large galaxy to our Milky Way, Andromeda (M31) is a spiral galaxy that is the most massive object in our Local Group of galaxies. It lies 2.5 million light-years away.

Angular momentum   A measure of the amount of rotation an object has, taking into effect its mass, shape, and speed.

Anisotropy   The property of being directionally dependent, as opposed to independent of directions; in cosmology, it refers to tiny differences in the cosmic microwave background radiation, which are clues to the early history of the universe.

Antimatter Material composed of antiparticles, particles with the same mass as normal matter but with opposite electrical charges.

Apollo missions The NASA program that sent 12 men to explore the Moon between 1969 and 1972.

Asteroid Small rocky body in the solar system, also called a minor planet.

Astrometry Positional astronomy that focuses on determining the precise positions of objects in the sky.

Astronomical unit The average distance between Earth and the Sun, defined as 149.598 million kilometers.

Astronomy **magazine** The world's largest publication about astronomy, headquartered near Milwaukee, Wisconsin, and edited by David J. Eicher since 2002.

Axion A hypothetical elementary particle proposed in 1977 as a possible component of cold dark matter.

B-mode polarization A polarization signal in the cosmic microwave background radiation that could help explain properties of the early universe.

Barred spiral galaxy A spiral galaxy with a prominent central bar of material emanating from the galactic center.

Baryonic matter Normal matter in the universe, composed of atoms, that comprises stars, planets, galaxies, and the luminous material we see in the cosmos.

BICEP2 A telescopic experiment being conducted at the South Pole, short for Background Imaging of Cosmic Extragalactic Polarization, that hopes to shed light on the nature of the early universe.

Big Bang The definitive cosmological model for the development of the universe, in which the cosmos is expanding and originated in a small space 13.8 billion years ago.

Big Bang nucleosynthesis The production of atomic nuclei heavier than hydrogen during the early history of the universe.

Big Crunch A possible scenario for the universe's fate, in which the expansion of space reverses and matter recollapses.

Big Freeze A highly likely scenario for the universe's fate, in which the universe expands forever, cools dramatically, and undergoes eventual heat death.

Big Rip A possible scenario for the universe's fate, in which matter in the cosmos is progressively torn apart by the universe's expansion.

Binary star A system of two stars orbiting a common center of mass.

Black hole A region of space-time in which intense gravity prevents anything, including light, from escaping.

Black hole era A hypothesized future period of the universe, 10^{40} to 10^{100} years from now, in which black holes will dominate the cosmos.

Boson In quantum mechanics, one of the two classes of particles, the other being fermions.

Cambrian explosion A short evolutionary event, beginning about 542 million years ago, in which enormous numbers of living species appeared.

Carbohydrate   A biological molecule consisting of carbon, hydrogen, and oxygen atoms, that performs numerous functions in living organisms.

Carbonaceous chondrite   A primitive class of meteorites that formed on the outer edge of the solar system and that contains primitive matter.

Cassini–Huygens mission   An ambitious spacecraft sent to explore Saturn and its moons, launched in 1997; the Cassini spacecraft entered orbit in 2004 and its Huygens lander touched down on Titan, one of Saturn's moons, in 2005.

CCD camera   A camera employing a charge-coupled device, an electronic photoreceptor that allows capturing a digital image with light-sensitive pixels (picture elements).

cD galaxy   A supergiant elliptical galaxy that is the central, dominant member of a galaxy cluster.

CDM   See Cold dark matter.

Central engine   In extragalactic astronomy, refers to the black hole in the center of an active galaxy.

Cepheid variable star   A star that varies between a larger, brighter state and a smaller, fainter one. The period and luminosity of these stars is so well known that they are used as "standard candles" in measuring distances in the cosmos.

Chandra X-ray Observatory   A space telescope launched in 1999 that has extensively studied the universe in the x-ray part of the spectrum.

Chandrasekhar limit   The maximum mass of a stable white dwarf star, about 1.4 times the mass of the Sun.

CHEOPS   CHaracterizing ExOPlanets Satellite, a European mission designed to study extrasolar planets, planned for a 2017 launch.

Chicxulub crater   A prehistoric impact crater buried underneath the Yucatán Peninsula in Mexico; created by the K-Pg Impact 66 million years ago.

Chirality   The property of asymmetry or "handedness" of molecules and other biologically significant factors.

Chondrite   Stony meteorites that have been modified due to heating or differentiation of the parent body, and which contain chondrules, rounded grains of minerals.

Civilization   A complex state society, marked with the advancement of life to a certain degree of intellectual and organizational sophistication.

CMB   See Cosmic microwave background radiation.

COBE   See Cosmic Background Explorer.

Cold dark matter   A hypothetical form of matter characterized by particles that move slowly relative to the speed of light.

Color-magnitude diagram   A modernized type of Hertzsprung–Russell diagram that plots a star's color index versus luminosity.

Comet   An icy small body of the solar system in orbit about the Sun.

Convection   The movement of molecules in fluids.

Corona   In planetary geology, an oval-shaped feature on Venus.

CoRoT   A French satellite, COnvection ROtation and Planetary Transits, designed to exoplanets with short periods and to conduct asteroseismology.

Cosmic Background Explorer A satellite launched in 1989 that provided evidence of the Big Bang formation of the universe.

Cosmic dark ages A period of the early universe in which the cosmos was opaque, approximately 150 million to 800 million years after the Big Bang.

Cosmic distance scale Distances in the universe; often also used in relation to the methods used to derive them.

Cosmic inflation See Inflation Theory.

Cosmic microwave background radiation Thermal radiation left over from the Big Bang origin of the universe; the relic radiation of the Big Bang.

Cosmic ray Incredibly powerful, high-energy radiation emanating from a variety of sources, mostly outside the solar system.

Cosmological constant The value of the energy density of the vacuum of space; denoted by Greek letter lambda, Λ.

Cosmology The study of the origin, evolution, and fate of the universe.

Cosmos Formally titled *Cosmos: A Personal Voyage*, Carl Sagan's 1980 PBS-TV series about the universe, and an accompanying book.

Cretaceous-Paleogene Extinction Event The K-Pg Extinction (for short, and also called the K-T Impact for Cretaceous-Tertiary, for the now outmoded term Tertiary) is a global extinction event that occurred about 66 million years ago and was caused by a roughly 10-kilometer asteroid plunging into what is now the Yucatán Peninsula in Mexico, creating Chicxulub Crater.

Cryosphere Portions of a planet or moon where water exists in solid form.

Curiosity A martian rover launched with the Mars Science Laboratory mission in 2011, which has been exploring Gale Crater on Mars since 2012.

Cyanobacteria A phylum of bacteria that obtain energy from photosynthesis.

Cygnus X-1 A well-known galactic black hole candidate discovered in 1964.

Dark energy A hypothetical form of energy believed to constitute 68.3 percent of the mass-energy in the universe; its existence resulted from analyses of distant supernovae in 1998.

Dark Era See Cosmic dark ages.

Dark matter A hypothetical type of matter believed to exist from its gravitational effects on bright matter, and probably constituting 26.8 percent of mass-energy in the cosmos.

Dark matter fluctuations Variations in the density of the cosmic microwave background radiation that provide evidence for dark matter.

Dark nebula A type of interstellar cloud, consisting of very fine dust grains, that obscures light from objects lying beyond it.

Darwinian evolution Also called Darwinism, a theory developed by Charles Darwin and others that explains the evolution of species over time; utilizes the principle of natural selection of small, inherited variations over time.

Deep Sky A quarterly publication about observing star clusters, nebulae, and galaxies edited by David J. Eicher that existed from 1982 through 1992.

Deep Sky Monthly A publication about observing star clusters, nebulae, and galaxies edited by David J. Eicher that existed from 1977 through 1982.

Deep-sky object   A star cluster, nebula, galaxy, or other object lying beyond the solar system.

Degenerate era   A hypothesized future time in the universe, some 10^{14} to 10^{40} years from now, in which star formation will completely cease, leaving stellar objects at a degenerate end.

Degenerate star   A compact star at the end of its evolutionary life that is not a black hole.

Density   An object's mass per unit volume.

DNA   Deoxyribonucleic acid, a complex organic molecule that encodes and contains the genetic instructions used in the development and function of living organisms.

Doppler shift   Also called the Doppler Effect, the change in frequency of a wave for an observer moving relative to the source.

Doppler technique   Also known as the radial velocity method, a form of detecting extrasolar planets by observing Doppler shifts in the spectrum of a star.

Drake Equation   A mathematical argument used to estimate the number of extraterrestrial civilizations in the Milky Way Galaxy.

Dwarf planet   A term adopted in 2006 by the International Astronomical Union to describe Pluto, Ceres, Haumea, Makemake, and Eris – solar system bodies in a solar orbit that are massive enough to have a gravitationally controlled shape but without having cleared their orbits of other celestial bodies.

Eccentricity   The amount by which a solar system body's orbit deviates from a circle.

Electromagnetic force   Also called electromagnetism, one of the four fundamental forces in the universe, described by electromagnetic fields.

Electron   A subatomic particle with a negative elementary electrical charge.

Elliptical galaxy   A galaxy with an ellipsoidal shape and a smooth brightness profile.

Embedding diagram   A diagram used to envision the properties of objects like black holes.

Emission nebula   A cloud of ionized gas glowing with high-energy photons and typically converting its gas into newborn stars.

Energy density   The amount of energy stored in a region of space per unit volume or mass.

Entropy   The measure of order in a thermodynamic system, which over time invariably moves from ordered to less ordered systems.

EPOXI   The renamed 2007-and-beyond phase of NASA's Deep Impact spacecraft.

European Space Agency   An intergovernmental agency founded in 1975, devoted to space exploration, with 20 member states.

Event horizon   In general relativity, a boundary of space-time beyond which events cannot affect an outside observer.

Exoplanet   See Extrasolar planet.

Extinction   The end of an organism or group of organisms, like a species.

Extrasolar planet   A planet orbiting a star other than the Sun, and therefore lying outside our solar system.

Faber–Jackson relation   An early empirical power law devised in 1976 by astronomers Sandra M. Faber and Robert E. Jackson, relating the luminosity and central stellar velocity dispersion of elliptical galaxies.

Faint early Sun paradox   Also called the Faint young Sun paradox, an apparent contradiction between observations that liquid water existed on early Earth and the belief that the young Sun emitted only 70 percent of its current radiation.

False vacuum   In quantum field theory, an apparently metastable region of space that appears to be in a vacuum state but is actually unstable.

Fusion   The nuclear process in which two or more atomic nuclei collide at high speed and join as a new nucleus, unleashing immense energy.

Gaia   A European Space Agency craft launched in 2013 to conduct high-precision mapping of 1 billion stars and other objects in the Milky Way Galaxy.

Galactic bar   The central bar-shaped structure in a barred spiral galaxy.

Galactic bulge   An enormous, tightly packed group of stars, gas, and dust around the center of a galaxy.

Galactic center   The rotational center of a galaxy.

Galactic cluster   See Open star cluster.

Galactic halo   An extensive, spherical component of a galaxy that envelops the primary, visible part of the galaxy.

Galactic plane   The plane in which the disk-shaped portion of a galaxy lies and rotates.

Galaxy   An enormous, gravitationally bound system of stars, gas, and dust.

Galaxy cluster   A structure bound by gravity and containing hundreds to thousands of individual galaxies.

Galaxy group   A structure made up of roughly 100 or less galaxies bound by gravity.

Galaxy supercluster   Large aggregates of galaxy groups and clusters, gravitationally bound, and forming the largest structures known in the universe.

General Theory of Relativity   A theory devised by Albert Einstein and presented in 1915 that describes geometric ideas about gravitation, and is the basis for gravitation in modern physics.

Giant Impact Hypothesis   The idea that the Moon formed from the collision of a Mars-sized body, Theia, with Earth about 4.5 billion years ago.

GLIMPSE   Galactic Legacy Infrared Mid-Plane Survey Extraordinaire, a survey made with the Spitzer Space Telescope that defined the structure of the Milky Way Galaxy, among other data.

Global warming   The observed, century-long rise in average temperatures in Earth's climatic system.

Globular star cluster   A spherical collection of hundreds of thousands to 1 million or more old stars bound by gravity, and orbiting a galaxy in its halo.

Gluon   Elementary particles that act as exchange particles for the strong force, between quarks.

Goldilocks Problem   The notion that Earth will not be habitable for living beings forever, perhaps for a billion years or less to come.

Gould's Belt   A partial ring of stars in the Milky Way stretching about 3,000 light-years across, and containing many bright O and B stars.

Grand Unified Theory   An imagined, sought-after theory in which the standard model of physics in which, at high energies, the gauge interactions of the four fundamental forces are all merged into a single force.

Gravitational force   One of the four fundamental forces in the universe, also known as gravity and gravitation; a natural phenomenon in the universe in which all bodies are attracted to other bodies.

Gravitational lensing   The effect when the gravity of a body bends light from a more distant body, forming two or more images, and thus acts as a lens.

Gravitational wave   Ripples in the curvature of space-time that act as propagating waves and are produced by black holes and other high-energy objects.

Gravity   See Gravitational force.

Great Attractor   A gravitational anomaly in space near the Hydra-Centaurus Supercluster, which lies within the Laniakea Supercluster, that betrays a concentration of mass many thousands of times greater than the Milky Way Galaxy.

Great Wall   Also called the CfA2 Great Wall or the Coma Wall, a superstructure in the universe discovered in 1989, measuring about 500 million light-years long, 300 million light-years wide, and 16 million light-years thick, and containing several superclusters of galaxies.

Greenhouse gas   An atmospheric gas that absorbs and emits infrared radiation.

HII region   A large, low-density cloud of gas that is partially ionized and within which infant stars are forming.

H-R diagram   See Hertzsprung–Russell diagram.

Habitable zone   Also called a "Goldilocks zone," the region around a star in which planets with sufficient atmospheric pressure would support liquid surface water.

Hadron   A composite particle made of quarks held together by the strong force.

Hadron era   The period of the early universe during which the cosmos was dominated by hadrons, from about 10^{-6} second to 1 second after the Big Bang.

Hawking radiation   Radiation predicted to be released from black holes due to quantum effects near the black hole's event horizon.

Heat death   A predicted cosmological event, suggested as the universe's ultimate fate, when the universe will no longer be able to sustain processes that consume energy. This may occur around 10^{1000} years from now.

Helioseismology   The study of wave oscillations inside the Sun.

Helium flash   Runaway fusion of helium in the core of a low-mass star or on the surface of a white dwarf star.

Hertzsprung–Russell diagram   A scatter graph of stars showing their luminosities versus their spectral classifications, first devised by Ejnar Hertzsprung and Henry Norris Russell around 1910.

Higgs boson   An elementary particle predicted by Peter Higgs and others in 1964 and likely confirmed in 2012, whose existence helps to explain why some particles have mass and why the weak force has a shorter range than electromagnetism.

High-z Supernova Search Team   An international collaboration of cosmologists using Type Ia supernovae to describe the expansion of the universe; in 1998, one of two teams involved in the discovery of dark energy.

Hipparcos   A European Space Agency satellite, launched in 1989, that operated until 1993, precisely measuring the positions of stars and numerous other celestial bodies.

Holocene   The current geological period, which began some 11,700 years ago, with the end of the Pleistocene.

Hot dark matter   A hypothetical form of dark matter involving particles that travel at velocities near the speed of light.

Hot Jupiter   A class of extrasolar planet similar to Jupiter in our solar system but with a range of high temperatures caused by orbiting close to their host stars.

Hubble constant   The estimated value of the rate of expansion of the universe, denoted with H_0. ESA's Planck mission data estimate the constant at 67.8 kilometers per second per megaparsec.

Hubble galaxy classification   Also called the Hubble Sequence, the scheme of galaxy classification devised by Edwin Hubble in 1936, describing spiral galaxies, barred spirals, and ellipticals.

Hubble Space Telescope   The most successful space telescope in history, launched in 1990 with a flawed mirror but then repaired in orbit, and contributing a vast wealth of knowledge about the cosmos.

Hubble Time   The age of the universe, 13.8 billion years, and the inverse of the Hubble constant.

Hubble Ultra Deep Field   An image of a small area of space in the southern constellation Fornax, made with the Hubble Space Telescope, showing extremely young and faint galaxies several hundred million years after the Big Bang.

Hydrothermal vent   A fissure in a planetary surface through which geothermally heated water escapes.

Hylomorphism   A philosophical theory derived by Aristotle, in ancient Greece, suggesting that living beings are a mixture of matter and form.

Inclination   In the orbits of astronomical objects, the angle of which one plane is oriented with respect to another.

Inflation Theory   A hypothetical, exponential expansion of the cosmos that occurred early in the history of the universe; developed in the 1980s by Alan Guth, Andrei Linde, and others.

Infrared Astronomical Satellite Also known as IRAS, a US satellite launched in 1983 that was the first to study the entire sky in the infrared part of the spectrum.

Infrared radiation Invisible radiant energy in the electromagnetic spectrum with longer wavelengths than visible light; most thermal (heat) radiation is infrared.

Inorganic matter Matter not containing carbon, although for historical reasons, some simple carbon-containing compounds are considered inorganic.

International Astronomical Union Abbreviated IAU, this international body of astronomers acts as the authority for naming and assigning designations to celestial bodies.

Interstellar medium The matter that exists in between stars in a galaxy, sometimes abbreviated ISM.

Interstellar space The space between stars and other objects in a galaxy.

Intragroup medium Superheated plasma present at the core of a galaxy group.

Ionization The process by which an atom or molecule acquires a charge, positively or negatively, by gaining or losing electrons, forming an ion.

Iron-Sulfur World A hypothetical early condition on Earth in which life may have originated on the surface of iron-sulfide minerals.

Irregular galaxy A galaxy without regular shape, unlike spiral, barred spiral, or elliptical galaxies.

Isotope Variants of chemical elements with different numbers of neutrons.

Jack Hills region A range of hills in Western Australia containing the oldest exposed surface rocks on Earth.

James Webb Space Telescope Abbreviated JWST, the space telescope planned for a 2018 launch that will succeed, in a sense, the Hubble and Spitzer space telescopes, idealized to study the universe in the infrared.

Jet Often called an astrophysical jet, a stream of matter observed in a variety of phenomena in astronomy, as with the centers of active galaxies, cataclysmic binary stars, x-ray binaries, and other objects.

JUICE A planned European Space Agency mission, Jupiter Icy Moon Explorer, which could be launched in 2022 to focus on Ganymede, Callisto, and Europa.

K-Pg Impact See Cretaceous-Paleogene Extinction Event

K-T Impact See Cretaceous-Paleogene Extinction Event

Kelvin The base unit of temperature measurement in science; 1 K = –272 °C or –458 °F.

Kepler Space Telescope An orbiting NASA observatory launched in 2009 that was remarkably successful in discovering extrasolar planets before losing partial function in 2013.

Kepler's laws Three basic laws of planetary motion devised by German mathematician Johannes Kepler between 1609 and 1619.

Kerr black hole A rotating, uncharged black hole with a spherical event horizon.

KREEP   An acronym for potassium (K), rare Earth elements (REE), and phosphorous (P), a component of lunar impact breccia and basaltic rocks.

Kuiper Airborne Observatory   A NASA-funded observatory placed into a Lockheed C-141A Starlifter jet that was commissioned in 1974 and flew until 1995, making observations of numerous solar system bodies in the infrared part of the spectrum. Named for Dutch–American astronomer Gerard P. Kuiper.

Kuiper Belt   A region of small solar system bodies beyond the planets, extending from the orbit of Neptune (at about 30 astronomical units from the Sun) to approximately 50 astronomical units from the Sun. Named for Dutch–American astronomer Gerard P. Kuiper.

Lagrangian point   Five positions in a planetary orbit such as Earth's where a small object affected by gravity can hold a stable position.

Lambda-CDM model   The standard cosmological model of the Big Bang in which the universe has a cosmological constant, associated with dark energy, and cold dark matter.

Large Hadron Collider   The world's largest particle collider, constructed by the European Organization for Nuclear Research (CERN) from 1998 to 2008, and located near Geneva, Switzerland.

Large Magellanic Cloud   A large irregular satellite galaxy of the Milky Way, lying 163,000 light-years away, and often abbreviated LMC.

Large-scale structure   Superclusters and filaments of galaxies, the largest known structures in the cosmos.

Late Devonian extinction event   One of five major extinction events in Earth's history, which took place about 374 million years ago.

Late Heavy Bombardment   A hypothesized period 4.1 to 3.8 billion years ago, during which Earth and other inner planets and moons underwent a large number of impacts from small bodies in the solar system.

Lenticular galaxy   A galaxy, intermediate in form, between spiral galaxies and elliptical galaxies, showing a lens shape, and with little active star formation.

Lepton era   A period of the universe from about 1 second to 10 seconds after the Big Bang, during which leptons dominated the mass of the universe.

Light curve   A graph showing the brightness of a celestial object versus time.

Light-year   The distance light travels in 1 year, approximately equal to 10 trillion kilometers (6 trillion miles).

Lipid   Naturally occurring molecules that include fats, waxes, sterols, triglycerides, and other substances.

Lipid World   A hypothesis about the origin of life on Earth suggesting that lipid-like structures were the first self-replicating molecules.

Local Group of galaxies   The local collection of galaxies near us in the cosmos, including the Milky Way, the Andromeda Galaxy, the Pinwheel Galaxy, and perhaps as many as 100 galaxies altogether.

Local Supercluster   Also called the Virgo Supercluster, the mass of galaxies in our region of the universe that contains the Local Group of galaxies, the Virgo Cluster, and at least 100 groups and clusters in a diameter of 110 million light-years.

LSP Lightest supersymmetric particle, the name given to the lightest hypothetical particle in models of supersymmetry; a dark matter candidate.

Luna missions A series of robotic lunar missions conducted by the Soviet Union between 1959 and 1976.

MACHO Massive compact halo object, a term for massive objects that might explain the presence of dark matter in galaxy halos; now believed unlikely to be the explanation for dark matter.

Magellan A NASA mission originally termed the Venus Radar Mapper, launched in 1989 and successful in mapping Venus by radar.

Magellanic Clouds Two satellite galaxies of the Milky Way, the Large Magellanic Cloud and the Small Magellanic Cloud.

Magnitude A logarithmic scale of measuring the brightnesses of astronomical objects.

Main Asteroid Belt The primary group of asteroids, lying between the orbits of Mars and Jupiter.

Main sequence A band of stars that appears diagonally across plots of stellar luminosity versus color, as with Hertzsprung–Russell diagrams.

Manhattan Project The US scientific project during World War II that produced the first atomic bombs.

Maria Large, dark basaltic plains on the Moon (singular, mare).

Mariner program A NASA program that sent interplanetary probes to Mars, Mercury, and Venus between 1962 and 1973.

Mars Climate Orbiter A NASA spacecraft launched toward Mars in 1998 and lost during its orbital insertion in 1999.

Mars Exploration Rover A space mission operated by NASA that sent two roving vehicles, Spirit and Opportunity, to Mars; Spirit landed at Gusev Crater and operated from 2004 through 2010, and Opportunity landed at Meridiani Planum in 2004 and continues science operations.

Mars Global Surveyor A NASA spacecraft launched in 1996 that operated in martian orbit through 2007.

Mars Pathfinder A NASA martian mission that landed on the Red Planet in 1997 with a base station named after Carl Sagan and a rover named Sojourner.

Mars Polar Lander/Deep Space 2 A NASA mission launched in 1999 to explore Mars, and which was lost during its orbital insertion.

Mars Reconnaissance Orbiter A NASA mission to Mars launched in 2005 that continues to operate in martian orbit.

Mars Science Laboratory A NASA robotic mission to Mars launched in 2011, which landed Curiosity, a large rover, in Gale Crater in 2012.

Mass extinction A widespread and rapid decrease in life on Earth.

Matter density If the density of matter, Ω_M, in the universe exceeds a critical value, the universe will be closed; if not, it will be open.

MAVEN Mars Atmosphere and Volatile EvolutioN, a NASA spacecraft launched in 2013 that will study the martian atmosphere.

Metallicity   The proportion of an astronomical object composed of elements heavier than hydrogen and helium.

Metal-poor   An astronomical object low in elements heavier than hydrogen and helium.

Meteorite   The physical fragment of an asteroid or comet that falls to Earth and is recovered; stony, iron, or a combination thereof.

Microfossil   Fossilized former organisms measuring less than 4 millimeters across.

Microquasar   A smaller, less energetic "cousin" of a quasar, composed of a radio emitting x-ray binary star.

Microwave background radiation   See Cosmic microwave background radiation.

Milkomeda   The name devised by astronomer Abraham Loeb for the eventual collision and merger product between the Milky Way and the Andromeda Galaxy.

Milky Way Galaxy   The barred spiral galaxy containing our Sun and solar system among its approximately 400 billion stars.

Millisecond pulsar   A pulsar, a highly magnetized, rotating neutron star, with a rotational period of about 1 to 10 milliseconds.

Molecular cloud   An interstellar cloud of gas and dust that supports the formation of molecules.

Molecular hydrogen   Hydrogen atoms existing in molecular form, as H_2, often associated with star formation.

Moon   A natural satellite of a planet.

The Moon   Earth's natural satellite, the largest in the solar system relative to its host planet.

MOST   Microvariability and Oscillations of STars, a Canadian space telescope launched in 2003 dedicated to the study of asteroseismology and extrasolar planets.

Multiple star   A star system composed of at least three suns, gravitationally bound.

Multiverse   A hypothetical set of other universes that could exist apart from our own; made plausible by mathematical models.

Muon   An elementary particle similar to an electron but with greater mass.

NASA   The US National Aeronautics and Space Agency, founded in 1958.

Near-Earth Asteroid   An asteroid whose orbit carries it into close proximity to Earth.

Near-Earth Object   An NEO is a solar system body whose orbit carries it into close proximity to Earth.

NEO   See Near-Earth Object

Neutrino   An electrically neutral, weakly interacting subatomic particle.

Neutron   A subatomic particle with no electrical charge and a mass slightly larger than a proton.

Neutron star   A stellar remnant that results from the collapse of a massive star following its supernova explosion.

New Horizons mission   A NASA space probe launched in 2006 that in 2015 became the first spacecraft to encounter Pluto and its system of moons.

NFW profile   Navarro–Frenk–White profile, a spatial mass distribution profile of dark matter attached to dark matter halos calculated by Julio Navarro, Carlos Frenk, and Simon White.

Noachian period   A geological system and early time period pertaining to the planet Mars, extending from about 4.1 billion to 3.7 billion years ago.

Nobel Prize   A set of annual awards for various sciences and for peace established in 1895 by Alfred Nobel and given by the government of Sweden.

Nucleic acid   Large biological molecules necessary for life that encode, express, and transmit genetic information.

Nucleocosmochronology   A technique to determine timescales associated with astronomical objects, employing abundances of radioactive nuclides to calculate ages.

Nucleon   One of the particles that makes up an atomic nucleus.

Occultation   A sky event that occurs when one object passes in front of another, hiding it from view.

Oort Cloud   A hypothesized shell of comets surrounding the solar system that may extend some 50,000 or more astronomical units, or nearly a light-year, away from the Sun – a quarter of the way to the nearest star. Named after Dutch astronomer Jan H. Oort.

Open star cluster   A group of several hundred to a few thousand young stars that formed from a molecular cloud and are gravitationally bound.

Opportunity   A NASA robotic rover active on Mars since 2004; part of the Mars Exploration Rover mission.

Orbital resonance   A phenomenon that occurs when two bodies exert a regular, periodic gravitational force, such that their orbital periods are related by a ratio of two integers.

Orbital velocity   The orbital speed of a body in a gravitational field.

Ordovician–Silurian extinction event   A range of events occurring about 450 to 440 million years ago that, taken together, were the second largest of the five major mass extinctions in Earth's history.

Organic matter   Matter composed of organic compounds containing carbon.

Orion Spur   A minor spiral arm or spur of the Milky Way Galaxy, some 10,000 light-years long and 3,500 light-years wide that contains the Sun and our solar system.

Parallax   The displacement of an object as viewed from two points of view; in astronomy, parallax observations are made when Earth is on opposite sides of the Sun.

Particle physics   The branch of physics that studies particles that make up matter and radiation.

Peculiar galaxy   A galaxy unusual in shape and form that defies classification as a spiral, barred spiral, elliptical, or irregular, and often shows chaotic form due to energetic events inside the galaxy.

Perihelion   The nearest point to the Sun in a solar system object's orbit.

Permian–Jurassic extinction event   One of several events known as the "Great Dying," a massive extinction that occurred some 252 million years ago.

Phantom energy   A hypothetical form of dark energy that would dramatically increase the expansion of the universe and cause the "Big Rip" end scenario.

Phoenix   A robotic NASA spacecraft launched in 2007 that studied Mars and its history of water in 2008.

Photon   An elementary particle that is the quantum carrier of light and all other forms of electromagnetic radiation and the force carrier of the electromagnetic force.

Photosynthesis   A process used by plants to convert energy from sunlight into chemical energy.

Pioneer Venus mission   A 1978 NASA mission that sent two spacecraft to study Venus.

Planck   A space observatory launched by the European Space Agency in 2009 that studied cosmological parameters through 2013.

Planck era   The earliest time of the chronology of the universe, from the Big Bang to approximately 10^{-43} second, when quantum effects of gravity dominated physical interactions.

Planck time   The time required for light to travel, in a vacuum, one Planck length, about 1.6×10^{-35} meters.

Planet   A spherical body, either gaseous or rocky, orbiting a star.

Planetary nebula   A type of emission nebula that results from the death of a star, with the approximate mass of the Sun, forming a transient, glowing cloud of gas.

Planetesimal   A protoplanet that is forming and growing larger by accretion in an early stage of its development, which is thought to have existed in large numbers in the early solar system and in other forming solar systems.

Plasma   Matter consisting of positively charged ions and negatively charged electrons.

Plate tectonics   A scientific theory that describes the large-scale motion of Earth's lithosphere, and operates or operated on other planets too.

PLATO   Planetary Transits and Oscillations of stars, a European Space Agency mission planned for launch in 2024, to study extrasolar planets.

Positron   An antimatter equivalent of the electron, with a positive charge.

Preplanetary nebula   A transitional nebular object resulting from a sunlike star's death that follows the asymptotic giant branch phase and precedes the planetary nebula stage.

Primordial black hole   A hypothetical black hole formed from the density of matter present in the universe's early expansion.

Principia Mathematica   Short for the full title, *Philosophiae Naturalis Principia Mathematica* (*Mathematical Principles of Natural Philosophy*), a landmark work produced by Isaac Newton and published in 1687.

Project Ozma   A pioneering SETI experiment created in 1960 by Frank Drake at the National Radio Astronomy Observatory in Green Bank, West Virginia.

Protein A large biological molecule consisting of one or more chains of amino acids.

Proton A subatomic particle with a positive electrical charge and a mass slightly less than a neutron.

Pulsar A highly magnetized, rotating neutron star.

Quantum cosmology A field attempting to determine the effects of quantum mechanics on the formation and evolution of the universe.

Quantum mechanics A branch of physics that deals with phenomena on nanoscopic scales.

Quantum physics Physics that is based on subatomic particles and the radiation associated with them.

Quantum theory See Quantum mechanics.

Quark An elementary particle that makes up hadrons, including protons and neutrons, the building blocks of atomic nuclei.

Quark theory The theoretical basis for differentiating and understanding types of quarks and how they behave.

Quasar The most energetic members of a class of objects known as active galactic nuclei; they represents the extremely powerful centers of young galaxies, driven by supermassive black holes.

Quintessence A hypothetical form of dark energy that was proposed to explain the accelerating expansion of the universe.

Radial velocity The velocity of an object relative to the direction of the radius; the line of sight between two objects.

Radial velocity method See Doppler technique.

Radiation density See Energy density.

Radiation pressure Electromagnetic radiation from the Sun pushing particles outward, away from the center of the solar system.

Radio galaxy A type of active galaxy that is very energetic in the radio portion of the electromagnetic spectrum.

Radiometric dating A technique used to date rocks or other ancient objects by measuring the amount of decay in radioactive isotopes they contain.

Rare Earth Hypothesis An idea that argues the emergence of complex life on Earth required an extremely unlikely series of events, and so complex extraterrestrial life is probably correspondingly rare.

Recombination In cosmology, the era when charged electrons and protons first bound together to form electrically neutral hydrogen atoms, some 378,000 years after the Big Bang.

Red dwarf A small, cool star on the main sequence; the most numerous type of star in the universe.

Red giant A luminous giant star, late in its lifetime, with a mass about one-third to 8 times that of the Sun.

Red Planet A casual name for the planet Mars.

Redshift A physical phenomenon caused when light or other electromagnetic radiation increases in wavelength, shifting toward the red end of the spectrum.

REE Rare Earth element, a set of 17 chemical elements.

Reflection nebula A cloud of interstellar dust composed of fine particles that reflect starlight, creating a dim, bluish nebula.

Refractory elements Dusty particles that have relatively high condensation temperatures, including metal and silicate grains.

Reionization The process that reionized matter following the cosmic dark ages, some 150 million to 1 billion years after the Big Bang.

Reproduction The biological process by which new offspring organisms are created.

Ridge belt A raised ridge on Venus suggesting evidence of shearing, compression, and other effects in early Venusian history.

RNA Ribonucleic acid, a family of large biological molecules that perform vital roles in the coding of genetic material.

RNA World A hypothesis suggesting that RNA molecules were the precursors to current life on Earth.

Rotation curve In galaxy research, a plot of the magnitude of orbital velocities of stars and gas versus their distance from the galactic center.

Rotational fission The notion that a satellite could separate from a parent body (such as the Moon from Earth) during early, very rapid rotation in the parent body's history.

Sagittarius A* The energetic radio source lying at the center of the Milky Way Galaxy, and associated with its central supermassive black hole, believed to have a mass of 4.1 million solar masses.

Sahelanthropus An extinct species of hominid, dating to 7 million years ago, about the time of the chimpanzee/human divergence.

Saros cycle A period of approximately 18 years and 11.3 days that can be used to predict eclipse occurrences.

Scattered Disk A distant region of the outer solar system that is sparsely populated by icy bodies – a subset of Trans-Neptunian objects.

Schwarzschild radius The radius of a sphere such that, if all mass of an object is compressed within the sphere, the escape speed from the sphere's surface would equal the speed of light.

Self-replicating molecules Molecules such as DNA that can build identical copies of themselves.

Semi-major axis One-half of the long axis of the ellipse of an orbit.

SETI The search for extraterrestrial intelligence, the collective name for a variety of scientific projects that commenced in 1960 and carry on today.

Singularity A point in space-time where quantities used to measure gravitation become infinite.

Sloan Digital Sky Survey A significant imaging redshift survey commenced in 2000, using a telescope at Apache Point, New Mexico.

Small Magellanic Cloud Abbreviated SMC, the smaller of two significant satellite galaxies of the Milky Way, lying at a distance of 197,000 light-years.

Sojourner The rover component of the Mars Pathfinder mission, launched in 1996 and completed in 1997.

Solar mass   A standard unit of mass in astronomy, equal to the mass of the Sun, about 1.99 x 10^{30} kilograms.

Solar system   Our Sun and its family of planets, moons, and small bodies.

Space-time   The mathematical model of understanding the universe that combines space and time into a continuum.

Special Theory of Relativity   A physical theory relating space and time, devised by Albert Einstein in 1905.

Spectral classification   Classification of stars based on their spectral types, from the hottest to coolest – O, B, A, F, G, K, and M.

Spectral line   A dark or bright line in an otherwise continuous spectrum, indicating the emission or absorption of light.

Spectrograph   An instrument that separates light into a frequency spectrum.

Spectroscopic parallax   A method for detecting stellar distances by comparing spectral types with apparent brightness.

Spectroscopy   The study of the interaction between light from an astronomical object and matter.

Spectrum   The record of an object's chemical properties in its emitted light.

Speed of light   A universal physical constant, and the fastest velocity in the universe; in a vacuum, it is 299,792,458 meters per second (186,282 miles per second), a quantity known as c.

Spiral galaxy   A galaxy with a distinct spiral pattern of arms emanating from its center.

"Spiral nebula"   A defunct name, attached to spiral galaxies before the nature of galaxies was discovered in the 1920s.

Spirit   A martian rover, part of the Mars Exploration Rover mission, that operated on Mars from 2004 through 2010.

Spitzer Space Telescope   A NASA infrared space observatory launched in 2003.

Standard candle   Physical distance indicators in the universe, such as Cepheid variable stars, Type Ia supernovae, and other objects that allow accurate distance determinations to objects that contain them.

Standard model   In particle physics, a theory involving the weak, strong, and electromagnetic forces, which govern the behavior of subatomic particles.

Star cluster   A group of stars bound together by gravity, and typically born together from a gas cloud. There are two main types, open and globular star clusters.

Stardust mission   A NASA spacecraft launched in 1999 that encountered and sampled Comet 81P/Wild 2 in 2004, returning samples for study in 2006.

Steady-state theory   A now-discredited idea about the origin and fate of the universe proposed by James Jeans, Fred Hoyle, and others in the 1920s through the 1940s.

Stellar association   A loose star group containing suns of a common origin but with some members that have become gravitationally unbound.

Stellar evolution   The study of how stars evolve through their lives.

Stelliferous era   The current era of the universe, which features, among other things, the formation of stars (beginning several hundred million years after the Big Bang). This era began about 1 million years after the Big Bang and will last until 100 trillion years from now.

Stratosphere   The second major layer of Earth's atmosphere, above the troposphere and below the mesosphere.

Strelley Pool   In Western Australia, an area that contains the earliest known microfossils of life, dating to about 3.4 billion years ago.

String theory   A hypothetical framework in particle physics in which point-like particles are replaced by one-dimensional strings.

Strong force   In particle physics, the strong nuclear force that ensures the stability of normal matter.

Subduction   The geological process by which one tectonic plate encounters the edge of another and slides underneath it, sinking into the mantle.

Sugar   A generalized name for sweet tasting, short-chain carbohydrates.

Sun   The star at the center of our solar system, one of approximately 400 billion in the Milky Way Galaxy and at least 10,000 billion billion in the universe.

Super Earth   An extrasolar planet with a mass greater than Earth's but substantially less than that of gas giants.

Supergiant star   Among the most massive and luminous stars, they have masses at least 8 times that of the Sun and luminosities a million times that of the Sun.

Supermassive black hole   A black hole existing at the center of a galaxy, with the mass of hundreds of thousands to billions of times greater than that of the Sun.

Supernova   A stellar explosion that can briefly outshine an entire galaxy, caused either by the sudden reignition of nuclear fusion in a degenerate star, or the gravitational collapse of the core of a massive star.

Supernova Cosmology Project   One of two research teams that discovered evidence for the acceleration of the expansion of the universe, and the existence of dark energy, in 1998.

Supersymmetry   In particle physics, the hypothetical extension of space-time symmetry that relates bosons and fermions.

SuperWASP   An extrasolar planet discovery project involving a large number of collaborators, and centered at Roque de los Muchachos Observatory on La Palma, Canary Islands; WASP stands for Wide Angle Search for Planets.

SWEEPS   Sagittarius Window Eclipsing Extrasolar Planet Search, an exoplanet project run jointly by NASA, ESA, and the Space Telescope Science Institute, using the Hubble Space Telescope, in 2006.

Terraforming   The hypothetical idea that one could "Earth shape" another planet or moon, making it habitable by humans.

TESS   Transiting Exoplanet Survey Satellite, a planned NASA space observatory scheduled for a 2017 launch.

Tessera   A complex, ridged surface geography on some plateau highlands on the planet Venus.

Theia   A hypothesized ancient planetesimal that, according to the Giant Impact Hypothesis, struck Earth some 4.53 billion years ago, creating the Moon. Named after the Greek goddess Theia, mother of Selene, the goddess of the Moon.

Thick disk A structural component of a galaxy, including the Milky Way, composed of relatively old stars and containing much of the densest part of the galaxy's bright disk portion.

Thin disk A structural component of a galaxy, including the Milky Way, composed of some 85 percent of the stars in the galactic plane, including nearly all young stars and with a wide range of stellar ages.

Time dilation In relativity theory, the time difference experienced by observers moving relative to each other or relative to gravitational masses.

Time warp In relativity theory, the warping of time around a mass in space-time.

Trans-Neptunian Object A TNO is an object in the solar system that orbits the Sun at a greater semimajor axis than that of Neptune, and includes many icy bodies such as comets.

Transit method The technique of discovering extrasolar planets by watching the light from host stars fade and brighten as unseen planets pass in front of them, partially blocking their light.

Triassic–Jurassic extinction event A mass extinction event on Earth 201 million years ago, marking the boundary between the Triassic and Jurassic periods.

Trigonometric parallax See Parallax.

Trojan asteroid An asteroid that shares an orbit with a planet or moon.

Tully–Fisher relation The relationship between the intrinsic brightness of a galaxy and the width of its emission lines, discovered by R. Brent Tully and J. Richard Fisher in 1977.

2MASS The Two Micron All-Sky Survey, a US project to map the entire sky in three infrared wavelengths, conducted from 1997 to 2003.

Type Ia supernova The explosion of a white dwarf star that reignites nuclear fusion either by accretion of material from another star or the merger of two white dwarfs; extremely useful because its precisely known brightness allows it to be a distance-measuring standard candle throughout the cosmos.

UFO An unidentified flying object – one of numerous lights or objects in the sky, mysterious to some observers, a few of whom mistake them for alien spacecraft!

ULIRG Ultra Luminous Infrared Galaxy, a rare type of galaxy with a luminosity of 10^{12} suns or more.

Ursa Major Moving Group A nearby stellar moving group, lying 80 light-years away, with a common motion and containing the bright stars of the Big Dipper.

Vega missions Soviet space probes that studied Comet Halley in the 1980s.

Venera program A series of probes sent to explore Venus by the Soviet Union between 1961 and 1984.

Venus Express The European Space Agency's first Venus mission, launched in 2005 and continuing to operate in venusian orbit.

Very Long Baseline Array Abbreviated VLBA, a system of radio telescopes operated from Socorro, New Mexico, using the technique of very long baseline interferometry.

Vesicle An assembly of lipid molecules, resembling a cell membrane.

Viking program A NASA Mars program that sent two probes, Viking 1 and Viking 2, to the Red Planet, two landers and two orbiters. They launched in 1975 and arrived in 1976, operating until 1982.

Virgo Cluster The closest large cluster of galaxies, lying some 55 million light-years away and containing at least 1,300 member galaxies.

Virgo Supercluster Also called the Local Supercluster, an enormous aggregation of at least 100 galaxy groups and clusters over a diameter of 110 million light-years; contains the Milky Way and our Local Group.

Vitalism The philosophy, espoused by Aristotle, that livings things are fundamentally different from non-living things because they contain non-physical elements.

VLBI Very-Long Baseline Interferometry, the radio astronomy technique of using multiple radio dishes spread around Earth to combine signals and emulate a telescope much larger than the individual dishes.

Volatiles Icy materials that have low boiling points and sublime away. They include elements and compounds such as water, ammonia, methane, carbon dioxide, nitrogen, and hydrogen.

Voyager program A NASA program to explore the major planets of the solar system in a "grand tour," with two spacecraft, Voyager 1 and Voyager 2. They were launched in 1977 and continue journeys through the outer solar system.

Wave/particle duality The concept that elementary particles exhibit the properties of both particles and waves.

Wavelength The length between successive crests of an electromagnetic wave.

Weak force In particle physics, the force responsible for the radioactive decay and nuclear fusion of subatomic particles.

White dwarf A degenerative stellar remnant, the end state for 97 percent of stars in the Milky Way, including the Sun; composed mostly of electron-degenerate matter.

Wilkinson Microwave Anisotropy Probe Abbreviated WMAP, a NASA cosmology probe launched in 2001 to study the cosmic microwave background radiation.

WIMP In cosmology, a weakly interacting massive particle, a hypothetical candidate for dark matter.

WMAP See Wilkinson Microwave Anisotropy Probe.

Wormhole Also known as an Einstein–Rosen Bridge, a hypothetical topological feature in space-time that could allow "shortcuts" through physical space.

X-ray binary A class of binary stars with high x-ray luminosity.

X-ray radiation A highly energetic form of electromagnetic radiation.

Bibliography

Alfaro, E. J. and A. J. Delgado, eds. *The Formation of the Milky Way*. New York: Cambridge University Press, 1995.

Amendola, Luca and Shinji Tsujikawa. *Dark Energy: Theory and Observations*. New York: Cambridge University Press, 2010.

Andrews, Bill, ed. *Explore the Solar System*. 2nd edn. Waukesha: Kalmbach Publishing Co., 2012.

Begelman, Mitchell and Martin Rees. *Gravity's Fatal Attraction: Black Holes and the Universe*. 2nd edn. New York: Cambridge University Press, 2010.

Bender, Ralf et al. "HST STIS Spectroscopy of the Triple Nucleus of M31: Two Nested Disks in Keplerian Rotation around a Supermassive Black Hole," *The Astrophysical Journal*, 631: 280–300, 2005.

Benjamin, Robert A. et al. "First GLIMPSE Results on the Stellar Structure of the Galaxy," *Astrophysical Journal Letters*, 630: L49, 2005.

Bennett, Jeffrey and Seth Shostak. *Life in the Universe*. 2nd edn. San Francisco: Pearson, 2007.

Berendzen, Richard, Richard Hart, and Daniel Seeley. *Man Discovers the Galaxies*. New York: Columbia University Press, 1984.

Bougher, S. W., D. M. Hunten, and R. J. Phillips, eds. *Venus II: Geology, Geophysics, Atmosphere, and Solar Wind Environment*. Tucson: University of Arizona Press, 1997.

Brake, Mark. *Alien Life Imagined: Communicating the Science and Culture of Astrobiology*. New York: Cambridge University Press, 2013.

Brown, Mike. *How I Killed Pluto and Why It Had It Coming*. New York: Spiegel & Grau, 2010.

Caldwell, Robert R., Marc Kamionkowski, and Nevin H. Weinberg, "Phantom Energy: Dark Energy with $w < -1$ Causes a Cosmic Doomsday," *Physical Review Letters*, 91: 2003.

Cameron, Alastair G. W. and William R. Ward. "The Origin of the Moon," *Abstracts of the Lunar and Planetary Science Conference*, 7: 120, 1976.

Canup, Robin M. "A Giant Impact Origin of Pluto–Charon," *Science*, 307: 546–550, 2005.

Canup, Robin M. and Larry W. Esposito. "Accretion of the Moon from an Impact-Generated Disk," *Icarus*, 119, 2: 427–446, 1996.

Canup, Robin M. and Erik Asphaug. "Origin of the Moon in a Giant Impact Near the End of the Earth's Formation," *Nature*, 412: 708–712, 2001.

"Simulations of a Late Lunar-Forming Impact," *Icarus*, 168, 2: 433–456, 2004.

Carr, Michael. *The Surface of Mars*. New York: Cambridge University Press, 2006.

Water on Mars. New York: Oxford University Press, 1996.

Cattermole, Peter. *Venus: The Geological Story*. Baltimore: The Johns Hopkins University Press, 1994.

Churchwell, Ed et al. *The Spitzer/GLIMPSE Surveys: A New View of the Milky Way*. *Publications of the Astronomical Society of the Pacific*, 121: 213 230, 2009.

Chyba, Christopher. "The New Search for Life in the Universe," *Astronomy*, 38: 5 (May 2010).

Cocconi, Giuseppe and Philip Morrison,"Searching for Interstellar Communications," *Nature*, 184: 844–846, 1959.

Coustenis, Athena and Thérése Encrenaz. *Life Beyond Earth: The Search for Habitable Worlds in the Universe*. New York: Cambridge University Press, 2013.

Cox, T. J. and Abraham Loeb. "The Collision between the Milky Way and Andromeda," *Monthly Notices of the Royal Astronomical Society*, 386: 461 (2008).

Eales, Stephen. *Origins: How the Planets, Stars, Galaxies, and the Universe Began*. London: Springer-Verlag, 2007.

Einstein, Albert. "Die Grundlage der allgemeinen Relativitätstheorie," *Annalen der Physik*, 49: 769–822, 1916.

"On a Stationary System with Spherical Symmetry Consisting of Many Gravitating Masses," *Annals of Mathematics*, 40: 922–936, 1939.

"Zur Elektrodynamik bewegter Körper," *Annalen der Physik*, 17: 891–921, 1905.

Ferris, Timothy. *Galaxies*. San Francisco: Stewart, Tabori, & Chang, 1982.

Freeman, Ken and Geoff McNamara. *In Search of Dark Matter*. Berlin: Springer, & Praxis Publishing, 2006.

Gargaud, Muriel, Purificación López-García, and Hervé Martin, eds. *Origins and Evolution of Life: An Astrobiological Perspective*. New York: Cambridge University Press, 2011.

Gates, Evalyn. *Einstein's Telescope: The Hunt for Dark Matter and Dark Energy in the Universe*. New York: W. W. Norton, 2009.

Gott, J. Richard, James E. Gunn, David N. Schramm, and Beatrice M. Tinsley,"An Unbound Universe," *The Astrophysical Journal*, 194: 543–553, 1974.

Gott, J. Richard and Robert J. Vanderbei. *Sizing Up the Universe*. Washington, DC: National Geographic Books, 2011.

Greenstein, George. *Understanding the Universe: An Inquiry Approach to Astronomy and the Nature of Scientific Research*. New York: Cambridge University Press, 2013.

Grinspoon, David Harry. *Venus Revealed: A New Look Below the Clouds of Our Mysterious Twin Planet*. Reading: Addison-Wesley, 1997.

Guth, Alan H.,"Inflationary Universe: A Possible Solution to the Horizon and Flatness Problems," *Physical Review D*, 23, 2: 347–356, 1981.

Harrison, Edward. *Cosmology: The Science of the Universe*. 2nd edn. New York: Cambridge University Press, 2000.

Hartmann, William K. and Donald R. Davis. "Satellite-Sized Planetesimals and Lunar Origin," *Icarus*, 24: 504–514, 1975.

Harwit, Martin. *Cosmic Discovery: The Search, Scope, and Heritage of Astronomy*. New York: Basic Books, 1981.

Hoskin, Michael, ed. *The Cambridge Illustrated History of Astronomy*. New York: Cambridge University Press, 1997.

Hubble, Edwin. *The Realm of the Nebulae*. New Haven: Yale University Press, 2013.

Huterer, Dragan and Michael S. Turner, "Prospects for Probing the Dark Energy via Supernova Distance Measurements," *Physical Review D*, 60: 1999.

Ida, Shigeru, Robin M. Canup, and Glen R. Stewart. "Lunar Accretion from an Impact-Generated Disk," *Nature*, 389: 353–357, 1997.

Impey, Chris. *How It Began: A Time-Traveler's Guide to the Universe*. New York: W. W. Norton, 2012.

   How It Ends: From You to the Universe. New York: W. W. Norton, 2010.

   "How Life Could Thrive on Hostile Worlds," in *Astronomy*, 36: 12 (December 2008).

Islam, J. N. *The Ultimate Fate of the Universe*. New York: Cambridge University Press, 2009.

Israelian, Garik, Brian May, and David Eicher, eds., *Starmus: 50 Years of Man in Space*. London: Canopus Publishing, 2014.

Jastrow, Robert and Michael Rampino. *Origins of Life in the Universe*. New York: Cambridge University Press, 2008.

Jayawardhana, Ray. *Strange New Worlds: The Search for Alien Planets*. Princeton, NJ: Princeton University Press, 2011.

Jones, Barrie W. *Pluto: Sentinel of the Outer Solar System*. New York: Cambridge University Press, 2010.

Kormendy, John, "Evidence for a Supermassive Black Hole in the Nuclus of M31," *The Astrophysical Journal*, 325: 128–141, 1988.

Kormendy, John and Luis C. Ho. *"Coevolution (Or Not) of Supermassive Black Holes and Host Galaxies,"* Annual Review of Astronomy and Astrophysics, 51: 511–653, Palo Alto: Annual Reviews, 2013.

Krauss, Lawrence M. and Michael S. Turner, "The Cosmological Constant is Back," *General Relativity and Gravitation*, 27: 1137–1144, 1995.

Kwok, Sun. *Cosmic Butterflies: The Colorful Mysteries of Planetary Nebulae*. New York: Cambridge University Press, 2001.

Lang, Kenneth R. *The Life and Death of Stars*. New York: Cambridge University Press, 2013.

Lankford, John, ed. *History of Astronomy: An Encyclopedia*. New York: Garland Publishing Co., 1997.

Lauer, Tod R. et al., "Planetary Camera Observations of the Double Nucleus of M31," *Astronomical Journal*, 106: 1436–1447, 1993.

Leakey, Richard. *The Origin of Humankind*. New York: Basic Books, 1994.

Leverington, David. *Encyclopedia of the History of Astronomy and Astrophysics*. New York: Cambridge University Press, 2013.

Lewin, Roger. *The Origin of Modern Humans*. New York: Scientific American Books, 1993.

Livio, Mario. *The Accelerating Universe: Infinite Expansion, the Cosmological Constant, and the Beauty of the Cosmos*. New York: John Wiley & Sons, 2000.

Livio, Mario, Megan Donahue, and Nino Panagia, eds. *The Extragalactic Distance Scale*. New York: Cambridge University Press, 1997.

Loeb, Abraham. *How Did the First Stars and Galaxies Form?* Princeton, NJ: Princeton University Press, 2010.

Loeb, Abraham and T. J. Cox. "Our Galaxy's Date with Destruction," in *The Milky Way*, Waukesha: Kalmbach Publishing Co., 2009.

Longair, Malcolm S. *The Origins of Our Universe: The Royal Institution Christmas Lectures*. New York: Cambridge University Press, 1991.

Lunine, Jonathan I. *Earth: Evolution of a Habitable World*. 2nd edn. New York: Cambridge University Press, 2013.

MacLeod, Norman. *The Great Extinctions: What Causes Them and How They Shape Life*. Richmond Hill, ON: Firefly Books, 2013.

Margulis, Lynn and Dorion Sagan. *What is Life?* New York: Simon & Schuster, 1995.

Marov, Mikhail Ya and David Grinspoon. *The Planet Venus*. New Haven: Yale University Press, 1998.

McFadden, Lucy Ann, Paul Weissman, and Torrence Johnson, eds. *Encyclopedia of the Solar System*. 2nd edn. Waltham: Academic Press, 2006.

Melia, Fulvio. *The Black Hole at the Center of Our Galaxy*. Princeton NJ: Princeton University Press, 2003.

  The Edge of Infinity: Supermassive Black Holes in the Universe. New York: Cambridge University Press, 2003.

Merloni, A., S. Nayakshin, and R. A. Sunyaev, eds. *Growing Black Holes: Accretion in a Cosmological Context*. Berlin: Springer, 2005.

Michell, John, "On the Means of Discovering the Distance, Magnitude, etc., of the Fixed Stars, in Consequence of the Dimension of the Velocity of Their Light, in Case Such a Diminution Should be Found to Take Place in Any of Them, and Such Other Data Should be Procured by Observations, as Would be Further Necessary for that Purpose," *Philosophical Transactions of the Royal Society of London*, 74: 35, 1783.

Newton, Isaac. *Philiosophiae Naturalis Principia Mathematica*. London: Samuel Pepys, 1687.

Novikov, Igor. *Black Holes and the Universe*. New York: Cambridge University Press, 1990.

Ostriker, Jeremiah P., Amos Yahil, and James E. Peebles, "The Size and Mas of Galaxies, and the Mass of the Universe," *The Astrophysical Journal*, 193: L1–L4, 1974.

Ostriker, Jeremiah P. and Paul J. Steinhardt, "The Observational Case for a Low-Density Universe with a Non-Zero Cosmological Constant," *Nature*, 377: 600–602, 1995.

Ostriker, Jeremiah P. and Simon Mitton. *Heart of Darkness: Unraveling the Mysteries of the Invisible Universe*. Princeton, NJ: Princeton University Press, 2013.

Peebles, P. James E., Lyman A. Page, and R. Bruce Partridge. *Finding the Big Bang*. New York: Cambridge University Press, New York, 2009.

Perlmutter, Saul et al. "Discovery of a Supernova Explosion at Half the Age of the Universe," *Nature*, 391: 51–54, 1998.

Perryman, Michael. *The Exoplanet Handbook*. New York: Cambridge University Press, 2011.

Rampino, Michael R. and Ken Caldeira. "The Goldilocks Problem: Climatic Evolution and Long-Term Habitability of Terrestrial Planets," *Annual Reviews of Astronomy and Astrophysics*, 32: 83–114, Palo Alto: Annual Reviews, 1994.

Rees, Martin. *Before the Beginning: Our Universe and Others*. Reading: Helix Books, 1997.
  Our Cosmic Habitat. Princeton, NJ: Princeton University Press, 2001.

Rowan-Robinson, Michael. *Cosmology*. 4th edn. New York: Oxford University Press, 2004.

Sanders, Robert H. *The Dark Matter Problem: A Historical Perspective*. New York: Cambridge University Press, 2010.

Scharf, Caleb. *Gravity's Engines: How Bubble-Blowing Black Holes Rule Galaxies, Stars, and Life in the Cosmos*. New York: Scientific American/Farrar, Straus & Giroux, 2012.

Seager, Sara, ed. *Exoplanets*. Tucson: University of Arizona Press, 2010.

Serjeant, Stephen. *Observational Cosmology*. New York: Cambridge University Press and The Open University, 2010.

Sohn, Sangmo Tony, Jay Anderson, and Roeland P. van der Marel, "The M31 Velocity Vector. I. Hubble Space Telescope Proper-Motion Measurements," *The Astrophysical Journal*, 753: 7–22, 2012.

Sparke, Linda S. and John S. Gallagher III. *Galaxies in the Universe: An Introduction*. 2nd edn. New York: Cambridge University Press, 2007.

Squyres, Steve. *Roving Mars: Spirit, Opportunity, and the Exploration of the Red Planet*. New York: Hyperion, 2005.

Stern, S. Alan. "On the Number of Planets in the Solar System: Evidence of a Substantial Population of 1000-km Bodies," *Icarus*, 90: 2, 1991.

Stern, S. Alan and David J. Tholen, eds. *Pluto and Charon*. Tucson: University of Arizona Press, 1997.

Stern, S. Alan and Jacqueline Mitton. *Pluto and Charon: Ice Worlds on the Ragged Edge of the Solar System*. New York: John Wiley & Sons, 1998.

Szczerba, Ryszard, Grażyna Stasińska, and Sławomir K. Górny, eds. *Planetary Nebulae as Astronomical Tools*. Melville: American Institute of Physics, 2005.

Talcott, Richard. "Seeking Ground Truth on Mars," *Astronomy*, 37: 10 (October 2009).

Thorne, Kip S. *Black Holes and Time Warps: Einstein's Outrageous Legacy*. New York: W. W. Norton, 1994.

Tremaine, Scott and James E. Gunn, "Dynamical Role of Light Neutral Leptons in Cosmology," *Physical Review Letters*, 42: 407–410, 1979.

Van den Bergh, Sidney. *Galaxy Morphology and Classification*. New York: Cambridge University Press, 1998.

Villard, Ray. "Skyfire: The Impending Birth of Our Supergalaxy," *Astronomy*, 41: 4 (April 2013).

Waller, William H. *The Milky Way: An Insider's Guide*. Princeton, NJ: Princeton University Press, 2013.

Xu, Y. et al. "On the Nature of the Local Spiral Arm of the Milky Way," *The Astrophysical Journal*, 769: 15, 2013.

Zwicky, Fritz, "On the Masses of Nebulae and Clusters of Nebulae," *The Astrophysical Journal*, 86: 217–246, 1937.

Index